GUIDE TO THE PRACTICAL USE OF CHEMICALS IN REFINERIES AND PIPELINES

GUIDE TO THE PRACTICAL USE OF CHEMICALS IN REFINERIES AND PIPELINES

JOHANNES FINK
Montanuniversitaet Leoben, Austria

ELSEVIER

AMSTERDAM • BOSTON • HEIDELBERG • LONDON
NEW YORK • OXFORD • PARIS • SAN DIEGO
SAN FRANCISCO • SINGAPORE • SYDNEY • TOKYO
Gulf Professional Publishing is an imprint of Elsevier

G | P
P |

Gulf Professional Publishing is an imprint of Elsevier
50 Hampshire Street, 5th Floor, Cambridge, MA 02139, USA
The Boulevard, Langford Lane, Kidlington, Oxford, OX5 1GB, UK

British Library Cataloguing in Publication Data
A catalogue record for this book is available from the British Library

Library of Congress Cataloging-in-Publication Data
A catalog record for this book is available from the Library of Congress

For information on all Gulf Professional publications visit our website at http://www.elsevier.com/

ISBN: 978-0-12-805412-3

Publisher: Joe Hayton
Acquisition Editor: Katie Hammon
Editorial Project Manager: Kattie Washington
Production Project Manager: Sruthi Satheesh
Designer: Greg Harris

CONTENTS

PREFACE

This book focuses on chemicals and processes for use in pipelines and the petroleum refining industry. The nature of the individual additives and the reasons why the individual additives act in the desired way are explained. The material presented here is compiled from the literature, including patents. In addition, as environmental aspects are gaining increasing importance, these issues are also dealt with.

HOW TO USE THIS BOOK

There are three indices: an index of acronyms, an index of chemicals, and a general index. In the chemical index, boldface page numbers refer to the sketches of structural formulas or to reaction equations.

J.K.F.

ACKNOWLEDGMENTS

The continuous interest and encouragement from Professor Wolfgang Kern, the head of the department of polymer chemistry, is highly appreciated. I am indebted to our university librarians, Dr Christian Hasenhüttl, Dr Johann Delanoy, Franz Jurek, Margit Keshmiri, Dolores Knabl, Friedrich Scheer, Christian Slamenik, and Renate Tschabuschnig, for support with literature acquisition. This book could not have been written otherwise. Thanks are given to Professor I. Lakatos, University of Miskolc, who directed my attention to this topic.

Finally, I thank the publisher for kind support, in particular Kattie Washington, Katie Hammon, and Sruthi Satheesh.

PART I

Pipeline Chemicals

CHAPTER 1

General Aspects of Pipelines

1.1 HISTORY

Pipelines have a long history. In ancient times, pipelines were used for water transport. Examples are still visible in archaeological areas. However, it is clear that these early constructions could not bear high pressures. The advent of gas pipelines occurred between 1820 and 1830 with the distribution of town gas. Nowadays, pipelines are indispensable in the petroleum industry for the transport of various materials, including natural gas, crude oil of various types, and refined products.

The construction and operation of pipelines are described in the literature [1–4], where the additives and chemicals that facilitate the transport of fossil fuel products are discussed.

1.2 MEDIA TO BE TRANSPORTED

1.2.1 Natural Gas

Properties of Natural Gas

Natural gas consists mainly of methane, although there are trace amounts of higher hydrocarbons, nitrogen, and even helium. It is typical in the gas transportation and storage industry to try to strip out higher hydrocarbons such as ethane, propane, butane, and unsaturated hydrocarbons from natural gas if the gas is to be transmitted through pipelines.

This leaves mostly methane (with some traces of nitrogen and carbon dioxide) to be transported by the gas pipeline. The materials that are stripped out are then transported or stored separately, often as liquids. A typical composition of natural gas transmitted through pipelines is shown in Table 1.1.

The compressibility factor z of methane is always less than 1.0 in the normal temperature range (ie, between −40°C and 50°C). Furthermore, the compressibility factor decreases as the pressure rises or the temperature

Guide to the Practical Use of Chemicals in Refineries and Pipelines
http://dx.doi.org/10.1016/B978-0-12-805412-3.00001-5

Table 1.1 Composition of the natural gas transmitted from Alberta, Canada, to Ontario, Canada

Component	Content (vol %)	Component	Content (vol %)
Methane	95.4	Ethane	1.97
Nitrogen	1.27	Carbon dioxide	0.55
Propane	0.51	Isobutane	0.17
n-Butane	0.08	Isopentane	0.02
n-Pentane	0.01	*n*-Hexane	0.02

Source: Morris I, Perry G. High pressure storage and transport of natural gas containing added C2 or C3, or ammonia, hydrogen fluoride or carbon monoxide. US Patent 6 217 626, assigned to JL Energy Transportation Inc. (Calgary, CA); 2001. http://www.freepatentsonline.com/6217626.html.

falls; therefore less energy is needed to pump a given volume of methane (measured at standard volume) at any given normal temperature than would be expected at that temperature if the methane behaved like an ideal gas.

This effect is more marked at higher pressures. Similarly, as the pressure is increased at a constant temperature, more methane (measured at standard volume) can be stored in a given volume than would be predicted from the ideal gas equation.

Below 7 MPa, the dominant variable for the compressibility factor in the *PVT* equation is the molecular weight of the gas. At this pressure, the addition of ethane or propane increases the molecular weight of the gas more rapidly than the compressibility factor z decreases it. Thus there is an advantage to removing ethane, propane, etc., from the gas.

At pressures greater than 7 MPa it is advantageous to add substances such as C_2 or C_3 hydrocarbon compounds, carbon monoxide, hydrogen fluoride, ammonia, or a mixture of these with natural gas. Ammonia without other additives is useful as an additive for gas storage at pressures down to about 5.5 MPa. Above a lower limit, which varies with the additive being added and the pressure, this results in a smaller $M_r \times z$ product, and therefore a decrease in the amount of power needed to compress the mixture for storage and to keep it compressed.

It is also advantageous to add ammonia to natural gas to be transmitted through pipelines at pressures above 5.5 MPa. Depending on the cost, it can also be advantageous to add carbon monoxide. Hydrogen fluoride is also effective, but is prohibitive because of its toxicity and corrosive properties.

One can find the precise amount of each additive that can be added at any pressure for beneficial results by calculating the product of the molecular weight and the compressibility factor z of the resulting mixture, $M_r \times z$, and comparing it with the product of the molecular weight and the compressibility factor z of the original natural gas. If $M_r \times z$ is smaller for the mixture than for the natural gas, energy can be saved in pumping and compression [5].

The use of two or more additives has a synergistic effect in many cases, so an even smaller amount of each is needed than would be the case if only one additive were present to produce the compressibility factor z over that of an equivalent standard volume of natural gas at the pressure and temperature involved.

When the mixture is pumped through a pipeline, however, an additional effect with ammonia emerges. In a pipeline there are pumping stations at intervals along the pipeline. At each pumping station the gas is compressed. As the gas moves toward the next pumping station, it gradually loses pressure and expands. The compression during passage through the compressor station heats the gas.

The gas cools while passing through the pipeline, transferring some of its heat to the surrounding soil through the pipeline wall. Ammonia has the property of being a refrigerant, which absorbs heat as it expands. Thus when a mixture of ammonia and natural gas is compressed and is then subsequently allowed to flow through a gas pipeline, the ammonia cools the mixture as it expands. This is regarded as an additional advantage [6].

Transportation Methods for Natural Gas

Natural gas has become a popular energy source worldwide. Natural gas is gaining in importance because of high oil prices. Traditionally, natural gas has been delivered to markets by two main methods [7]: (1) by pipelines and (2) as liquefied natural gas.

Each of these methods is capital intensive. Most advanced technologies include onshore gas-to-liquid floating liquefied natural gas, natural-gas hydrates, adsorbed natural gas, and compressed natural gas.

Alternatives to compressed natural gas technology, the operating differences between liquefied natural gas technology and compressed natural gas technology, and a technical and economic evaluation have been presented [7].

In both liquefied natural gas technology and compressed natural gas technology the gases are stored forms of natural gas. However, the key difference is that compressed natural gas is a gas that is stored at high pressure, while liquefied natural gas is an uncompressed liquid form that is stored at a low temperature.

Thus compressed natural gas requires a much larger volume to store the same mass of natural gas in comparison with liquefied natural gas. The following are used for the transportation of compressed natural gas:

- large wound coils of pipe,
- cargo-containment cylinders,
- tank modules for compressed natural gas,
- composite-reinforced pressure vessels,
- pressure vessels made of fiber-reinforced–plastic.

All these methods are economically comparable. In practice, the methods for the transportation of liquefied natural gas can transport more than twice the amount of gas that could be transported by the methods for the transportation of compressed natural gas. This is because of the increased weight of the storage containers for compressed natural gas, which must be made thicker to withstand the high pressure.

On the other hand, the processing of compressed natural gas is easier than that of liquefied natural gas. The transportation chain for liquefied natural gas starts with gas treatment, liquefaction, and storage before loading of the carriers for liquefied natural gas. This requires a comparatively complicated loading system. Eventually after transport, the liquid is pumped into pipelines of a vaporization system. In the case of compressed natural gas, the transportation chain includes only dehydration and compression at the loading terminal.

It has been concluded that compressed natural gas is very attractive for small reserves. Further, it requires a smaller total investment [7].

1.2.2 Blending of Hydrogen into Natural Gas

Blending hydrogen into an existing natural gas pipeline network has been proposed as a means of increasing the output of renewable energy systems such as large wind farms [8]. If this is implemented with relatively low concentrations, less than 5–15% hydrogen by volume, this strategy of storing and delivering renewable energy to markets appears to be viable without significantly increasing risks associated with use of the gas blend in end-use devices, such as household appliances, overall public safety, or the durability and integrity of the existing natural gas pipeline network.

The impact of hydrogen blends on industrial facilities must be studied in detail. Stationary gas engines will likely require changes to their control systems [9]. Three major risks of adding hydrogen to natural gas are:

1. gas buildup,
2. explosions in enclosures,
3. risk from transmission pipelines.

Compared with natural gas transmission pipeline explosions, there is a consistent tendency for the severity of the risk with hydrogen mixtures to shift spatially, increasing closer to the point of explosion and decreasing further from the point of explosion [10].

1.2.3 Crude Oil Blends

It is has been predicted that after 2014, crude oil production from the Alaska North Slope will decline so much that there will be a need to pump additional liquids through the pipeline to maintain the economical operation of the Trans-Alaska Pipeline System [11].

One of the proposed modes of transportation of the gas-to-liquid products is commingling them with crude oil so that it will from a single phase and then pumping the mixture through the pipeline. This transportation mode changes the properties of the gas-to-liquid product as well as the crude oil.

The physical and chemical properties of the gas-to-liquid product and its blends with crude oil from the Alaska North Slope have been assessed. From the data, the optimum blend ratio to minimize the transportation costs economics has been evaluated. It has been demonstrated that the flow behavior of the gas liquid blends is very temperature sensitive.

The viscosity and the density of the blends were found to decrease with increasing amounts of gas liquid and with increasing temperature. The rheograms for the specific crude oil are shown in Fig. 1.1. The flow behavior is Newtonian at temperatures above 20°C and changes to Bingham below 20°C.

The optimum gas-to-liquid ratio in the blend is obtained from the average pressure drop per mile under the specific conditions of transport. This is shown in Fig. 1.2.

The optimum amount of gas to liquid in the blend is governed by various factors, such as flow behavior parameters of the fluid, pipeline characteristics, elevation changes, and operation temperature. The optimum gas-to-liquid blend to crude oil blend ratio was found to be 1:2.5 [11].

1.2.4 Heavy Crude Oils

Heavy crude oils constitute a significant portion of the known global petroleum reserves [12]. Vast deposits of heavy crude oils are located in

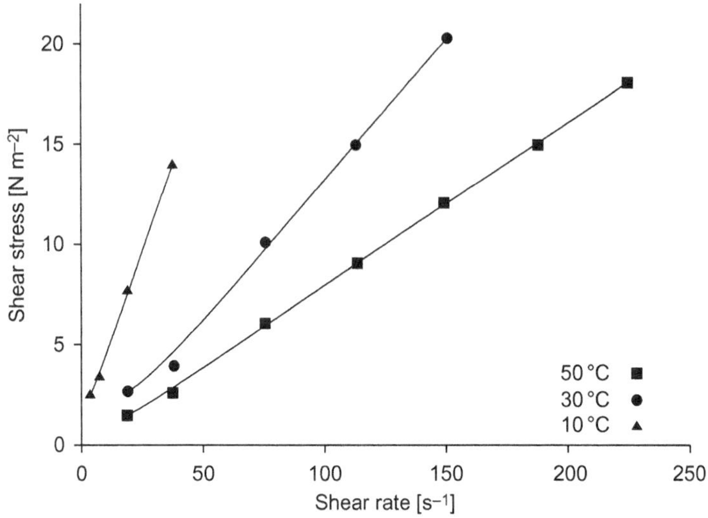

Figure 1.1 Rheograms for crude oil at various temperatures. *(Source: Abhijeet IA, Chukwu GA, Khataniar S, Patil SL, Dandekar AY. Rheology of gas-to-liquid products: Alaska north slope (ANS) crude oil and their blends for transportation through the trans-Alaska pipeline system (TAPS). SPE Proj Facilit Construct 2006; 1(4):118. http://dx.doi.org/10.2118/100375-PA).*

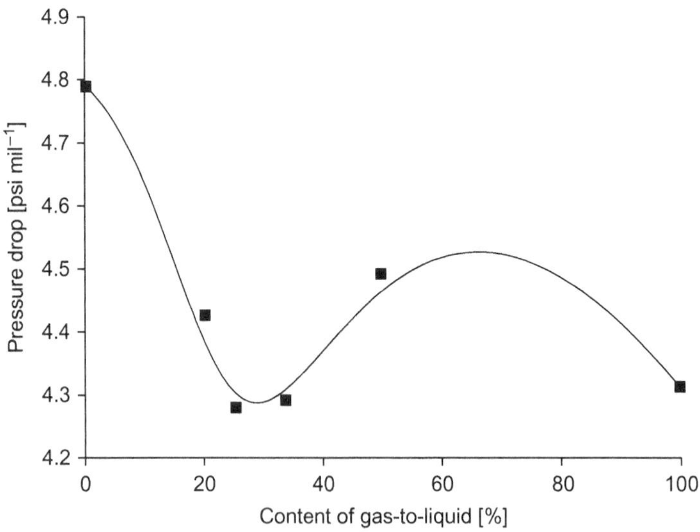

Figure 1.2 Pressure drop versus gas-to-liquid ratio. *(Source: Abhijeet IA, Chukwu GA, Khataniar S, Patil SL, Dandekar AY. Rheology of gas-to-liquid products: Alaska north slope (ANS) crude oil and their blends for transportation through the trans-Alaska pipeline system TAPS. SPE Proj Facilit Construct 2006; 1(4):118. http://dx.doi.org/10.2118/100375-PA).*

Colombia, Venezuela, Mexico, and Canada. Heavy crude oils are highly viscous or solid at ambient temperature, and have an American Petroleum Institute (API) gravity of 20 or less. Heavy crude oils include the high molecular weight hydrocarbons referred to as tars, petroleum tars, or tar sands, such as the deposit known as the Athabasca Tar Sands in Alberta, Canada.

The production of heavy crude oil and bitumen is very costly. This arises because of the extremely low mobility caused by high viscosity [13].

In addition of being more viscous than conventional crude oils, some heavy crude oils are rich in asphaltenes, metals, and resins. Some heavy crude oils contain high levels of sulfur, nitrogen, nickel, and vanadium [14]. Further, they are rich in the condensed polyaromatic compounds which react readily to form coke. The presence of these types of compounds may result in various problems in the recovery, transportation, treatment, and refining of these oil types [12].

On the other hand, the transportation of heavy and extra heavy crude oils from the well head to the refinery is becoming important because their production is currently rising all over the world. Such oils are characterized by an API gravity of less than 20 and high viscosity. The issues of transportation of heavy and extra heavy crude oil by pipelines have been reviewed [15].

Various methods have been developed to transport heavy oils through pipelines. The commonest method is to blend the heavy crude oil with a light hydrocarbon to decrease the overall viscosity. Different technologies for reduction of the viscosity have been reviewed as have the advantages and disadvantages [14]. The methods for facilitating the transport of heavy crude oils are summarized in Table 1.2.

Table 1.2 Easier transport of heavy crude oils

Viscosity reduction	Drag reduction	In situ upgrading
Dilution or blending Heating Emulsification Pour point reduction	Drag reduction additives Core annular flow	Catalytic upgrading Thermal degradation

Source: Hart A. A review of technologies for transporting heavy crude oil and bitumen via pipelines. J Petrol Explorat Product Tech 2013;4(3):32736. http://dx.doi.org/10.1007/s13202-013-0086-6.

The catalytic upgrading process in situ was specifically developed for the recovery and upgrading of heavy oil and bitumen [16]. The effects of the reaction gas media used in the upgrading process were investigated, in particular the effects of hydrogen, methane, nitrogen, and a blended gas mixture. The viscosity reduction and the conversion of hydrocarbons into lower boiling fractions follows the same trend as the API gravity.

The most relevant parameters in pipeline transportation of heavy crude oil are velocity, viscosity, temperature, density, and pour point [17]. Heavy crude can be transported on trunk systems in a variety of modes, including segregation, blending, and batching.

Segregation requires separate pipelines, blending consists of mixing crude oils, and batching refers to shipping crude in discrete batches. There are a number of methods for enhancing the transportability of heavy crude. These include oil-in-water emulsion formation, droplet suspension, dilution, the use of drag-reducing additives, and heating.

Another method is to establish a two-phase flow. This results in pipeline lubrication. A thin water film is injected around the internal oil core, which leads to the core annular flow regime. Thus the water lubricates the heavy oil, and the longitudinal pressure gradient is largely reduced. This basic method is more than 100 years old and goes back to an early US patent [18]. A sketch from that patent is shown in Fig. 1.3.

In Fig. 1.3, *A* represents a pipeline through which the fluid desired to be conveyed is caused to flow. The flow may be occasioned by the pump *B*, which draws the fluid by a suction pipe *b* from a suitable source. With the first fluid delivered to the pipeline *A*, a second fluid, which must be of greater specific gravity than the first fluid, may be delivered to the first fluid in the pipeline by means of the pump *C*, with its suction *c*, encircling sleeve *c* and holes c' in the pipeline *A*.

Experiments with such systems revealed a pressure-drop reduction of more than 90% compared with the same product without lubrication. These results confirm the effectiveness of such a lubricating process [13].

1.2.5 Emulsions

Water-external emulsions can be transported in pipelines to achieve higher net flow rates of oil than in the transport of dry oil alone [19]. The percentage of oil in water may range between 70% and 80%. The oil is combined with an aqueous solution comprising water, a pH-enhancing agent, and solid particles and is mixed until the solids-stabilized oil-in-water

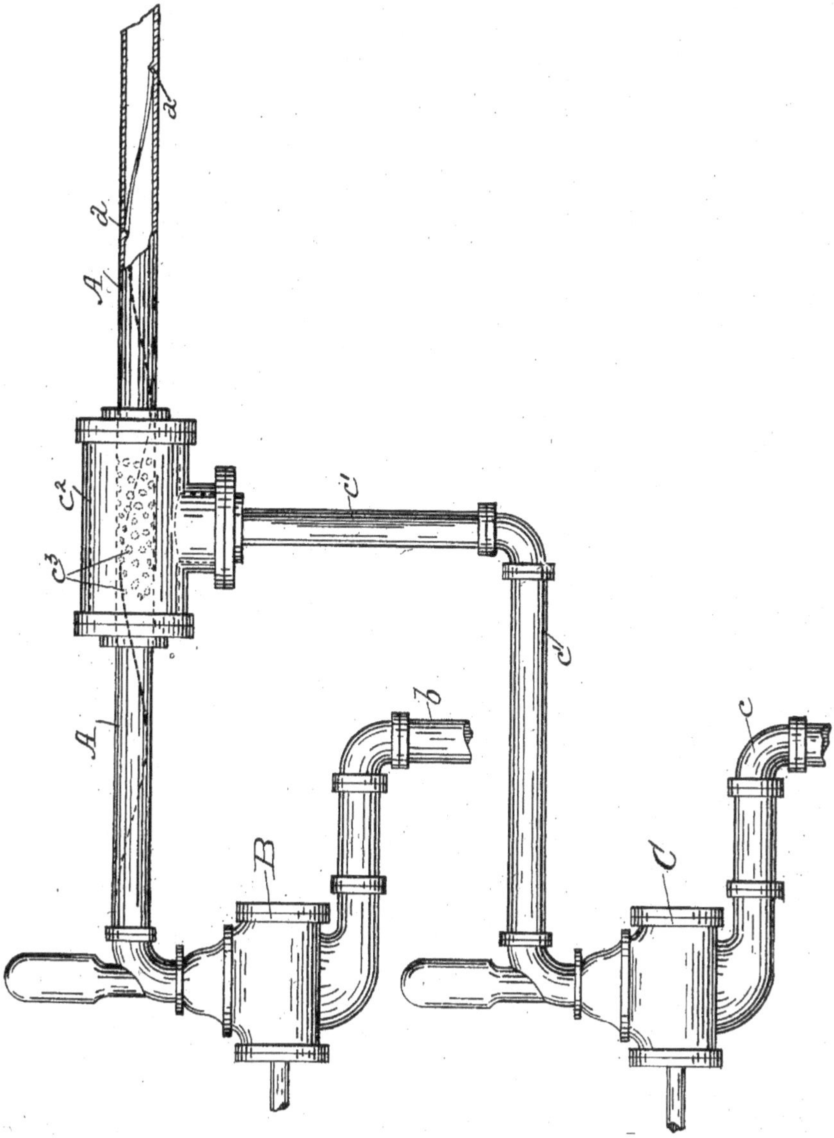

Figure 1.3 Piping method.

emulsion is formed. The pH of the resulting oil-in-water emulsion should be 7.5–10.

In the propagation of the emulsion through a pipe it is preferred to contact the inner walls of the pipe first with a wettability-altering agent to make the inner walls of the pipe water-wet so as to facilitate the propagation of the oil-in-water emulsion. After the inner walls of the pipe have been contacted with the wettability-altering agent, the oil-in-water emulsion can be pumped through the pipe.

High-oil content, solids-stabilized oil-in-water emulsions are therefore good candidates for transportation in pipelines using flow regimes of either self-lubricating core annular flow or uniform, lower-viscosity water-external emulsions [19]. In core annular flow, the formation of a low-viscosity annulus near the pipe wall further reduces pressure drop.

The viscosity of water (ie, the continuous phase) is not greatly affected by temperature. Therefore the viscosity of a solids-stabilized oil-in-water emulsion is not greatly affected by temperature and such oil-in-water emulsions do not have to be heated to high temperatures to maintain an acceptably low viscosity for economical transport.

Another benefit of oil-in-water emulsions is that the oil phase does not tend to wet the steel. Thus these emulsions will have fewer tendencies to wet or foul the pipeline walls than oil-external emulsions or dry oil [19].

Freezing Point Depressants

Mixtures of aqueous emulsions of oil can be more effectively transported through pipelines if certain antifreeze formulations are added to the system. Stable oil-in-water emulsions for pipeline transmission by use of 0.05–4% ethoxylated alkyl phenol as an emulsifier and a freezing point depressant for water enable pipeline transmission at temperatures below the freezing point of water [20].

Highly viscous petroleum oil containing 30–80% water can be transported through pipes more efficiently when a 1:1 mixture of washing liquid and antifreeze (ie, ethylene glycol with borax) is added to the oil in amounts of 0.002–0.2%. In addition to the increased efficiency of transport, reduced corrosion of pipes can be achieved [21].

1.2.6 Slurry Transport

The history of pipeline transport of high-density slurries has been reviewed. Shortcomings in the past and current technologies have been

Table 1.3 Materials for slurry pipeline transport

Compounds
Bauxite pumping
Gold tailings
Coal slurries
Limestone pipelines
Copper slurry
Iron ore slurry
Phosphate and phosphoric acid slurries
Clay and drilling muds
Oil sands
Uranium tailings

Source: Abulnaga B. Slurry systems handbook. New York: McGraw-Hill; 2002. ISBN 9781591243915.

described [22]. The necessary equipment and the construction of pipelines have been reviewed [23].

Slurry transport uses several basic principles [24]:

- hydraulic conveying,
- conventional lean slurry conveying,
- high-concentration slurry disposal.

For oil sand plants, a mixture of oil sand and water must often be pumped over a long distance to release the bitumen by ablation. These pipelines are also called hydrotransport pipelines. Materials that can be transported in slurry pipelines are summarized in Table 1.3.

Examples of additive uses are the slurry pipeline transport of minerals, the removal of the solids produced during the drilling of wells, and the removal of solids formed during the polishing and grinding of metals. Anionic, cationic, or nonionic surfactants may be used to alter the viscosity—for example, cetyltrimethylammonium chloride, dodecyl diphenyl oxide disulfonate, and soya bis(2-hydroxyethyl)amine [25].

Fatty amines are prepared from the corresponding fats by both conversion into the nitrile and hydrogenation. Soybean oil contains a mixture of various long-chain acids. The major acids are oleic acid (9-octadecenoic acid) and linoleic acid (9,12-octadecadienoic acid)—that is, C_{18} single and double unsaturated acids, respectively. Components for additives for slurry transport are shown in Fig. 1.4. Fatty amines are highly active surfactants.

A method to transport soft clay in a pipeline has been described [26]. This method should overcome the difficulties of the transport of soft clay

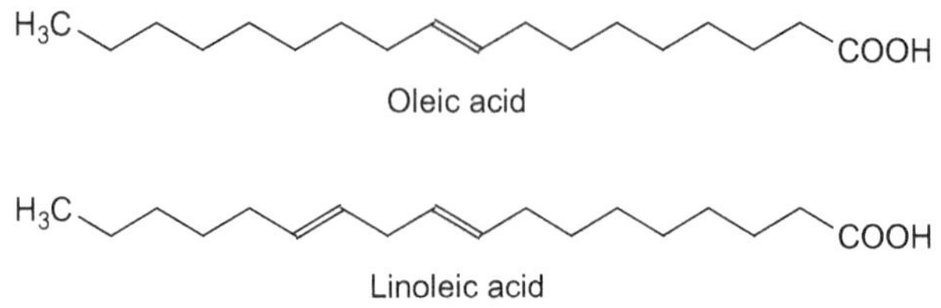

Figure 1.4 Components for additives for slurry transport.

so that transportation over a long distance is less problematic. By analysis of the properties of soft clay and pipeline transportation, the rules to decrease the energy through the pipeline transportation, and the characters of flow field with gas, liquid, and mud, a mathematical model for the pipeline transportation of soft clay has been developed.

1.3 TESTING AND DESIGN OF PIPELINES

The integrity of a pipeline is very important in many industrial and engineering systems. The aspects related to pipeline integrity have been reviewed and analyzed [27]. Design practices for pipelines are discussed on the basis of pressure criteria [28]. Inspection techniques have also been reviewed. These are the basis of integrity assessment techniques. Suggestions for the further development of automated tools to support pipeline integrity management have been presented. Subsequently, specific aspects of this field are described.

The specific issues related to the design of pipelines for installation by horizontal directional drilling have been described in detail [29]. These pipelines can be installed by medium-sized to large horizontal directional drilling rigs.

1.3.1 Blockage Detection in Natural Gas Pipelines

Pipeline blockage may become a severe problem in gas transportation [30]. A simple numerical model has been developed for accurate blockage characterization in natural gas pipelines. The transport phenomenon is modeled with a set of partial differential equations for isothermal natural gas flow in the pipeline. Transient pressure–wave reflection analysis is used to describe the blocking phenomenon. Reflected pressure waves that are observed at the pipe inlet node are then analyzed for the characterization of a blockage phenomenon. Viscous losses have no effect on the blockage length and the accuracy of the assessment of the location.

1.3.2 Maintenance Models for Petroleum Pipelines

A model has been applied to derive the optimum maintenance strategy for a petroleum pipeline. The model incorporates structured expert judgment to estimate the frequency of failure, taking into account various failure mechanisms [31].

Optimization models are applied to derive optimum maintenance intervals for petroleum pipelines on the basis of the frequency of failure estimated.

Two separate maintenance-optimization models have been proposed, a use-based optimization model that minimizes the expected total cost from a petroleum pipeline and a benefit-cost ratio model that maximizes the benefit from the pipeline, while minimizing operation and failure costs.

As failure mechanisms, external interference, corrosion, structural defect, and operational errors have been taken into consideration. Details of the mathematical calculation procedures have been presented [31].

The testing, comparison, and application of global sensitivity techniques for the study of the impact of the stream impurities on CO_2 pipeline failure have been described [32]. Global sensitivity analysis through nonintrusive generalized polynomial chaos expansion with sparse grids has been compared with more commonly used techniques and has been found to achieve a superior convergence rate in comparison with crude Monte Carlo and quasi-Monte Carlo methods for functions with up to a moderate level of roughness.

The simulation method has been described in detail. The method has been applied to a hypothetical full-bore rupture of a carbon dioxide pipeline. The results of the study suggest that the outflow rate is highly sensitive to the composition during the early stages of pressure reduction, where the effect of the impurities on phase equilibria has a significant impact on the outflow [32].

1.3.3 Interfacial Rheological Properties

The interfacial rheological properties are extremely sensitive to the chemical composition of formulations of immiscible formation liquids [33]. Therefore comparison and interpretation of the interfacial rheological properties may contribute significantly to extension of the spectrum of reservoir characterization, better understanding of the displacement mechanism, development of more profitable enhanced and improved oil recovery methods, intensification of surface technologies, optimization of pipeline transportation, and improvement of refinery operations [34].

When asphaltenes are precipitated, one can remove them from the walls of a well or a pipeline by washing them with a hydrocarbon solvent. It has been shown that isopropyl benzoate is exceptionally useful as a solvent for asphaltene removal [35].

1.3.4 Optimal Design for Gas Transmission Pipelines

Usually, natural gas transmission pipelines transport large quantities of natural gas over long distances [36]. They operate at high pressures and use a series of compressor stations at frequent intervals along the pipeline to move the gas over long distances.

On the basis of mass and momentum balance, a rigorous analytical equation has been derived for the compressible fluid flow in pipelines [37].

For a pipeline with constant cross-sectional area

$$\frac{\partial \rho v}{\partial x} = 0, \tag{1.1}$$

where ρv is the product of the gas density and gas velocity, which is constant according to Eq. (1.1). x is the flow direction along the pipeline axis. The density is calculated from the ideal gas equation. The energy equation for the gas flow is

$$\frac{\partial \rho v^2}{\partial x} = \frac{\partial p}{\partial x} - \frac{f \rho v^2}{2d} - \rho g \sin \alpha, \tag{1.2}$$

where f is the friction factor, g is the Earth acceleration, and α is the angle of deviation from the horizontal direction. The friction factor can be calculated as [38]

$$\frac{1}{\sqrt{f}} = 2 \log \left\{ \frac{\varepsilon / d}{3.7065} - \frac{5.0452}{N_{Re}} \log \left[\frac{1}{2.8257} \left(\frac{\varepsilon}{d} \right)^{1.1098} + \frac{5.8506}{N_{Re}^{0.8981}} \right] \right\}, \tag{1.3}$$

where $N_{Re} = \frac{\rho v d}{\mu}$, where μ is the gas viscosity.

The equation expresses the relationship among the flow rate, inlet pressure, and outlet pressure. This relationship is useful for design calculations.

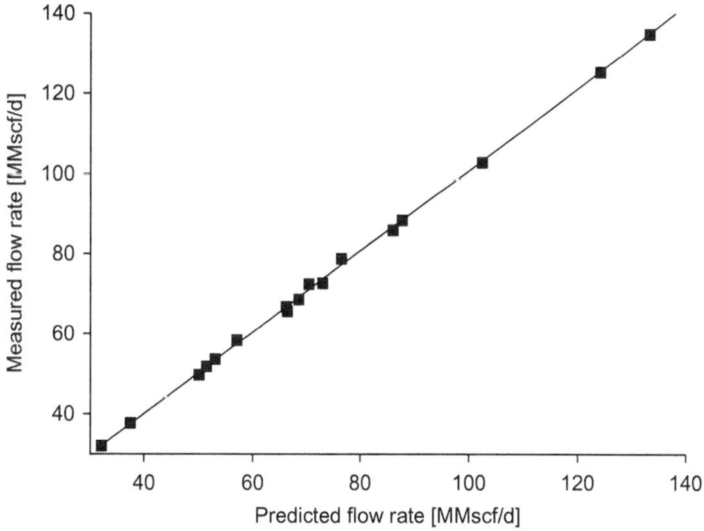

Figure 1.5 Predicted flow rate and field data. *MMscf*, Million standard cubic feet. *(Source: Tian S, Adewumi MA. Development of analytical design equation for gas pipelines. SPE Product Facil 1994;9(02):1006. http://dx.doi.org/10.2118/24861-PA).*

In particular, the equation can be used for pipeline topology and configuration. Computer programs have been developed to handle such applications. Problems that have been studied with this equation include bottom hole pressure calculations, gas injection calculations, and long-distance gas pipeline design calculations. Excellent agreement between the predicted results and specific field data has been found [37]. Differences of only 0.5–1% were observed. A comparison of predicted flow rate and field data is shown in Fig. 1.5.

Most of the gas in China is transported from well heads to markets via cross-country gas networks. The network comprises four large-diameter pipelines and will include most major gas pipelines.

A study aimed at optimizing the network to minimize its energy consumption and cost has been presented [39]. The large size and complex geometry of the network demands the study be broken down into simple components:

1. optimizing the operation of the components locally,
2. recombining the optimized components into the network,
3. optimizing the network globally.

This approach uses four different optimization methods: a penalty function method, pattern search, enumeration, and nonsequential dynamic programming. The results revealed that the cost savings, because of global optimization, are reduced with increased throughput [39].

The optimization problem has been solved with use of a number of decision variables and a number of constraints. The objective function includes [36]:

- installation costs,
- compressor stations,
- fuel consumption in compressor stations,
- maintenance,
- labor,
- supervision.

A software computer program [40] has been used to obtain the solution procedure for optimal design and operation of gas transmission pipelines [36].

1.3.5 Selection of Pipeline Routes

The selection of pipeline routes requires detailed knowledge of the seabed conditions and the shallow geological features. The areas along the routes and the areas regionally around the routes must be accurately surveyed. Specific geohazards may be [41]:

- steep slopes,
- unstable slopes,
- hard seabed sediments,
- bedforms,
- environmentally sensitive areas.

Environmentally sensitive areas are biological assemblages of tapeworms, clams, mussels, and bacterial mats. Further, a variety of organisms that are collectively known as chemosynthetic communities deserve attention.

These issues have been proved to be especially important for the complex geology and seabed topography found in the deepwater Gulf of Mexico.

In the Mardi Gras project, an autonomous underwater vehicle was used to collect high-quality survey data. These data allowed the project to make decisions with a high degree of confidence with regard to pipeline routing. Thus the Mardi Gras pipelines could be successfully installed [41].

1.4 STANDARDS

For convenience, some technical standards for the characterization of materials that can be used for piping are given subsequently.

1.4.1 Density

Accurate determination of the density, relative density, or API gravity of petroleum and its products is necessary for the conversion of measured volumes to volumes or masses, or both, at the standard reference temperature of 15°C. Density measurement can be done by ASTM D1298, a standard test method for density, relative density, or API gravity of crude petroleum and liquid petroleum products by the hydrometer method [42]. The API gravity is a factor governing the quality and pricing of crude petroleum.

1.4.2 Viscosity

Many petroleum products are used as lubricants, and the correct operation of the equipment depends on the appropriate viscosity of the liquid being used. In addition, the viscosity of many petroleum fuels is important for the estimation of optimum storage, handling, and operational conditions. Thus the accurate determination of viscosity is essential to many product specifications. The viscosity is measured by ASTM D445 [43].

1.4.3 Pour Point

The pour point of a petroleum sample is an index of the lowest temperature of its utility for certain applications. The pour point is measured by ASTM D97 [44].

1.4.4 Sulfur Content

The quality of many petroleum products is related to the amount of sulfur present. Knowledge of sulfur concentration is necessary for processing purposes. There are also regulations promulgated by federal, state, and local agencies that restrict the amount of sulfur present in some fuels.

A method for rapid and precise measurement of total sulfur in petroleum and petroleum products with a minimum of sample preparation has been developed [45]. A typical analysis time is 1–5 min per sample.

The analytical techniques for sulfur determination in crude oil, fuels and lubricants, coal, oil shale, gaseous fuels, and biofuels have been summarized [46].

1.4.5 Boiling Range Distribution

The boiling range distribution of petroleum fractions is, among other factors, dependent on the paraffin content [47]. This provides insight into the composition of petroleum feedstocks and products. Gas chromatography can be used to replace the conventional distillation methods.

1.4.6 Carbon Number Distribution

The IP 372 standard is a test method for the quantitative determination of the carbon number distribution of distillate paraffin waxes in the carbon number range from 18 to 40 [48].

1.4.7 Corrosion

The petroleum production environment provides a unique challenge for corrosion testing as it is not a chemical process environment, but is an environment provided by nature, where all the potential variability may occur [49]. The corrosion test program for petroleum environments must account for the environment nature has created, with very little ability to alter or control it.

Observations and data derived from corrosion testing can be used to determine the average rate of corrosion that occurs during an exposure interval [50].

REFERENCES

[1] Mohitpour M, Golshan H, Murray A. Pipeline design & construction: a practical approach. New York: American Society of Mechanical Engineers; 2000.
[2] Bourgoyne Jr AT., Keith KM, Martin EC, Young Jr FS. Applied drilling engineering. SPE Textbook Series, vol. 2. Richardson, TX: SPE Publications; 1986.
[3] Krass W, Kittel A, Uhde A. Pipelinetechnik: mineralölfernleitungen. Cologne: Verlag TÜV; 1979.
[4] McAllister EW, editor. Pipeline rules of thumb handbook: quick and accurate solutions to your everyday pipeline problems. 7th ed. Burlington: Gulf Professional Publishing, Elsevier Inc.;2009.
[5] Morris I, Perry G. Pipeline transmission method. US Patent 6 201 163, assigned to JL Energy Transportation Inc. (Calgary, CA); 2001. http://www.freepatentsonline.com/6201163.html.
[6] Morris I, Perry G. High pressure storage and transport of natural gas containing added C2 or C3, or ammonia, hydrogen fluoride or carbon monoxide. US Patent 6 217 626, assigned to JL Energy Transportation Inc. (Calgary, CA); 2001. http://www.freepatentsonline.com/6217626.html.
[7] Nassar YM. Comparisons and advantages of marine CNG transportation. SPE Proj Facilit Construct 2010;5(4):225–9. http://dx.doi.org/10.2118/127974-PA, doi: 10.2118/127974-pa.

[8] Melaina MW, Antonia O, Penev M. Blending hydrogen into natural gas pipeline networks: a review of key issues. Review of Key Issues NREL/TP-5600-51995. Golden, CO: National Renewable Energy Laboratory; 2013. https://www.eere.energy.gov/hydrogenandfuelcells/pdfs/blending_h2_nat_gas_pipeline.pdf.
[9] Florisson O. Preparing for the hydrogen economy by using the existing natural gas system as a catalyst. NaturalHy Project Report SES6/CT/2004/502661. Groningen, The Netherlands: Nederlandse Gasunie; 2010. https://www1.eere.energy.gov/hydrogenandfuelcells/pdfs/05_florisson.pdf.
[10] Lowesmith BJ, Mumby C, Hankinson G, Puttock JS. Vented confined explosions involving methane/hydrogen mixtures. Int J Hydrogen Energy 2011;36(3):2337–43. The Third Annual International Conference on Hydrogen Safety, http://www.sciencedirect.com/science/article/pii/S0360319910003848, http://dx.doi.org/10.1016/j.ijhydene.2010.02.084.
[11] Abhijeet IA, Chukwu GA, Khataniar S, Patil SL, Dandekar AY. Rheology of gas-to-liquid products: Alaska north slope (ANS) crude oil and their blends for transportation through the trans-Alaska pipeline system (TAPS). SPE Proj Facilit Construct 2006;1(4):1–18. http://dx.doi.org/10.2118/100375-PA.
[12] Chheda BD, Banavali RM, Mazza G. Recovery and transportation of heavy crude oils. US Patent 6 402 934, assigned to Rohm and Haas Company (Philadelphia, PA); 2002. http://www.freepatentsonline.com/6402934.html.
[13] Peysson Y, Bensakhria A, Antonini G, Argillier JF. Pipeline lubrication of heavy oil: experimental investigation of flow and restart problems. SPE Product Operat 2007;22(1). http://dx.doi.org/10.2118/97764-PA.
[14] Hart A. A review of technologies for transporting heavy crude oil and bitumen via pipelines.J Petrol Explorat Product Tech 2013;4(3):327–36. http://dx.doi.org/10.1007/s13202-013-0086-6.
[15] Martínez-Palou R, de Lourdes Mosqueira M, Zapata-Rendán B, Mar-Juárez E, Bernal-Huicochea C, de la Cruz Clavel-López J, et al. Transportation of heavy and extra-heavy crude oil by pipeline: a review. J Petrol Sci Eng 2011;75(3–4):274–82. http://www.sciencedirect.com/science/article/pii/S0920410510002640, http://dx.doi.org/10.1016/j.petrol.2010.11.020.
[16] Hart A, Leeke G, Greaves M, Wood J. Down-hole heavy crude oil upgrading by CAPRI: effect of hydrogen and methane gases upon upgrading and coke formation. Fuel 2014;119:226–35. http://www.sciencedirect.com/science/article/pii/S0016236113011198, http://dx.doi.org/10.1016/j.fuel.2013.11.048.
[17] Gerez JM, Pick AR. Heavy oil transportation by pipeline. In: Proceedings Volume, vol. 2, 1st ASME et al international pipeline conference (Calgary, Can, 6/9–13/96); 1996. p. 699–710. ISBN 0-7918-1767-9.
[18] Isaacs JD, Speed JB. Method of piping fluids. US Patent 759 374, assigned to John Dove Isaacs and James Buckner Speed; 1904. http://worldwide.espacenet.com/publicationDetails/biblio?DB=worldwide.espacenet.com&II=0&ND=3&adjacent=true&locale=en_EP&FT=D&date=19040510&CC=US&NR=759374A&KC=A.
[19] Bragg JR, Varadaraj R. Solids-stabilized oil-in-water emulsion and a method for preparing same. US Patent 7 121 339, assigned to ExxonMobil Upstream Research Company (Houston, TX); 2006. http://www.freepatentsonline.com/7121339.html.
[20] Gregoli AA, Olah AM. Low-temperature pipeline emulsion transportation enhancement. US Patent 5 156 652, assigned to Canadian Occidental Petroleum Ltd. (Calgary, CA); 1992. http://www.freepatentsonline.com/5156652.html.
[21] Van Neste WA, Maes JP, Darden JW, Triebel CA. Corrosion-inhibited antifreeze formulation. US Patent 4 759 864, assigned to Texaco Inc. & S.A. Texaco Petro, N.V. (White Plains, NY); 1988. http://www.freepatentsonline.com/4759864.html.

[22] Paterson AJC. Pipeline transport of high density slurries: a historical review of past mistakes, lessons learned and current technologies. Mining Tech 2012;121(1):37–45. http://dx.doi.org/10.1179/1743286311Y.0000000020.
[23] Abulnaga B. Slurry systems handbook. New York: McGraw-Hill; 2002. ISBN 9781591243915.
[24] Wikipedia. Slurry transport—wikipedia, the free encyclopedia. 2013. [Online; accessed 26-August-2014], http://en.wikipedia.org/w/index.php?title=Slurry_transport&oldid=561382479.
[25] Teot AS, Rose GD, Stevens GA. Friction reduction using a viscoelastic surfactant. US Patent 4 615 825, assigned to The Dow Chemical Company (Midland, MI); 1986. http://www.freepatentsonline.com/4615825.html.
[26] Zhu WZ, Wu SH, Tao SL. A mathematical model for the pipeline transportation of soft clay.Appl Mech Mater 2013;275–277:1476–9. http://dx.doi.org/10.4028/www.scientific.net/AMM.275-277.1476.
[27] Kishawy HA, Gabbar HA. Review of pipeline integrity management practices. Int J Pressure Vessels Piping 2010;87(7):373–80. http://www.sciencedirect.com/science/article/pii/S0308016110000827, http://dx.doi.org/10.1016/j.ijpvp.2010.04.003.
[28] Guo B. Offshore pipelines design, installation, and maintenance. 2nd ed. Waltham, MA: Gulf Professional Publishing; 2014. ISBN 9780123984920.
[29] Skonberg E, Muindi TM, editors. Pipeline design for installation by horizontal directional drilling. 2nd ed. Reston, Virginia: American Society of Civil Engineers; 2014. ISBN 9780784478370.
[30] Adeleke N, Ityokumbul MT, Adewumi M. Blockage detection and characterization in natural gas pipelines by transient pressure-wave reflection analysis. SPE J 2012;18(2):355–65. http://dx.doi.org/10.2118/160926-PA.
[31] Dawotola A, van Gelder PHAJM, Vrijling JK. Integrity maintenance of petroleum pipelines. Oil Gas Facil 2012;1(6):40–45. http://dx.doi.org/10.2118/162873-PA.
[32] Brown S, Beck J, Mahgerefteh H, Fraga ES. Global sensitivity analysis of the impact of impurities on CO_2 pipeline failure. Reliabil Eng Syst Safety 2013;115: 43–54. http://www.sciencedirect.com/science/article/pii/S0951832013000410, http://dx.doi.org/10.1016/j.ress.2013.02.006.
[33] Lakatos I, Lakatos-Szabo J. Effect of IOR/EOR chemicals on interfacial rheological properties of crude oil/water systems. In: SPE international symposium on oilfield chemistry, Society of Petroleum Engineers; 2001. p. 1–10. http://dx.doi.org/10.2118/65391-MS.
[34] Lakatos-Szabó J, Lakatos I, Kosztin B. Role of interfacial rheological properties of oil/water systems in mechanism and design of EOR/IOR technologies. In: Proceedings Volume, no. 057. 9th EAGE Impr. Oil Recovery Europe Symp. (The Hague, Netherlands, 10/20–22/97) Proc; 1997. ISBN 90-73781-08-6.
[35] Cox T, Grainger N, Scovell EG. Maintenance of oil production and refining equipment. EP Patent 1 276 828 assigned to Imperial Chemical Industries PLC; 2003. https://www.google.at/patents/EP1276828A1?cl=en.
[36] Elshiekh TM, A Khalil S, El Mawgoud HA. Optimal design and operation of Egyptian gas-transmission pipelines. Oil Gas Facil 2013;2(4):44–8. http://dx.doi.org/10.2118/163087-PA.
[37] Tian S, Adewumi MA. Development of analytical design equation for gas pipelines. SPE Product Facil 1994;9(02):100–6. http://dx.doi.org/10.2118/24861-PA.
[38] Chen NH. An explicit equation for friction factor in pipe. Indust Eng Chem Fundament 1979;18(3):296–7. http://dx.doi.org/10.1021/i160071a019.
[39] Jin L, Wojtanowicz AK. Optimization of large gas pipeline network—a case study in China. J Can Petrol Tech 2010;49(04):36–43. http://dx.doi.org/10.2118/136345-PA.

[40] LINDO Systems. Lingo 3.1 modeling software. Internet; 1997. http://lingo.software.informer.com/13.0/.
[41] Tootill NP, Vandenbossche MP, Morrison ML. Advances in deepwater and ultradeep-water pipeline route selection. SPE Proj Facil Construct 2006;1(1):1–4. http://dx.doi.org/10.2118/103640-PA.
[42] -1B. Relative density, or API gravity of crude petroleum and liquid petroleum products by hydrometer method. ASTM Standard, Book of Standards, Vol. 05.01 ASTM D1298-1B, ASTM International, West Conshohocken, PA; 2012. http://dx.doi.org/10.1520/D1298-12B.
[43] ASTM D445. Standard test method for kinematic viscosity of transparent and opaque liquids (and calculation of dynamic viscosity). ASTM Standard, Book of Standards, Vol. 05.02 ASTM D445-12, ASTM International, West Conshohocken, PA; 2012. http://dx.doi.org/10.1520/D0445-12.
[44] ASTM D97. Standard test method for pour point of petroleum products. ASTM Standard, Book of Standards, Vol 05.01 ASTM D97, ASTM International, West Conshohocken, PA; 2012. http://dx.doi.org/10.1520/D0097-12.
[45] ASTM 4294. Standard test method for sulfur in petroleum and petroleum products by energy dispersive X-ray fluorescence spectrometry. ASTM Standard, Book of Standards, Vol. 05.02 ASTM 4294, ASTM International, West Conshohocken, PA; 2010. http://dx.doi.org/10.1520/D4294-10.
[46] Nadkarni KRA. Sulfur: chemical analysis of fossil fuel products; 2014. http://dx.doi.org/10.1520/MONO11-EB.
[47] ASTM D2887. Standard test method for boiling range distribution of petroleum fractions by gas chromatography. ASTM Standard, Book of Standards, Vol. 05.01 ASTM D2887, ASTM International, West Conshohocken, PA; 2013. http://dx.doi.org/10.1520/D2887.
[48] IP 372. Determination of carbon number distribution of paraffin wax—gas chromatography method. London IP 372, Energy Institute, West Conshohocken, PA; 1985. http://www.energypublishing.org.
[49] Baboian R, Hack HP, Efird KD. Petroleum. In: Baboian R, editor, Corrosion tests and standards: application and interpretation, 2nd ed. West Conshohocken, PA: ASTM International; 2005. p. 425–34 [chapter 64]. ISBN 9780803120983, http://dx.doi.org/10.1520/MNL11042M.
[50] ASTM G4. Standard guide for conducting corrosion tests in field applications. ASTM Standard, Book of Standards, Vol. 03.02 ASTM G4, ASTM International, West Conshohocken, PA; 2008. http://dx.doi.org/10.1520/G0004-01R08.

CHAPTER 2

Leakage in Pipelines

An integrated quantitative risk assessment method for systematically analyzing alternatives for protecting oil and gas pipelines and pipeline freight transportation has been presented in a monograph [1]. That book analyzes engineering, environmental, and economic impacts and addresses both direct and collateral damage. Also regulations, standard processes, and risk assessment models recommended by the US Department of Homeland Security were presented. Statistical data on leakages in pipelines are given in Table 2.1.

2.1 SIMULATION METHODS

An efficient method for the analysis of corroding pipelines has been developed [2]. The procedure uses the first-order reliability method and system reliability approaches to evaluate the time-dependent failure probabilities of a pressurized pipeline at a single active corrosion defect. Three different failure modes are taken into consideration [2]:

1. small leak,
2. large leak,
3. rupture.

The criteria for distinguishing a small leak, a large leak, and a rupture at a given corrosion defect have been established from information available in the literature.

Wedge integral and probability weighting factor methods are used to evaluate the probabilities of a small leak and burst, whereas the conditional reliability index method is used to evaluate the probabilities of a large leak and a rupture.

The proposed method can be used to facilitate reliability-based corrosion management programs for energy pipelines [2].

Guide to the Practical Use of Chemicals in Refineries and Pipelines
http://dx.doi.org/10.1016/B978-0-12-805412-3.00002-7

Table 2.1 Leakages in pipelines

Reason	Amount (%)
Corrosion	36.42
Other	21.60
Excavation	15.39
Natural force	8.47
Material defect	6.98
Equipment	6.57
Operation	2.53
Other outside force	1.86

Source: Melaina MW, Antonia O, Penev M. Blending hydrogen into natural gas pipeline networks: a review of key issues. Review of Key Issues NREL/TP-5600-51995, National Renewable Energy Laboratory, Golden, CO, 2013. ttps://www.eere.energy.gov/hydrogenandfuelcells/pdfs/blendingh2natgaspipeline.pdf.

2.2 LEAKAGE OF POLYMER SUPPORTS

Hydrogen is more mobile than methane in many polymer materials, including the plastic pipes and elastomeric seals used in natural gas distribution systems [3]. This occurs because the permeation coefficient of hydrogen is higher through most elastomeric sealing materials than through plastic pipe materials. However, pipes have much larger surface areas than seals, so leaks through plastic pipe walls would account for the majority of gas losses. In typical polymer pipes, the permeation rates of hydrogen are about four to five times higher than those of methane.

A calculation based on literature data for the permeation coefficient of hydrogen and methane in polyethylene pipes suggested that most gas loss would occur through the pipe wall rather than through the joints in distribution mains smaller than 2 in that are operating at 5 bar or higher.

Gas leakage from elastomeric seals at joints in service lines may increase the risk in confined spaces, and this topic demands an additional risk assessment.

The investigation into specific pipe and seal materials and systems can provide a basis to estimate gas leakage more accurately. This basis can be used to determine whether leakage in confined spaces might present a safety risk over time and the degree to which detection and monitoring devices may be required to manage risks [3].

2.2.1 Failure of Elastomers

Failure of elastomers could result in leaks. The major failure of elastomers comes from the chemical reaction between the elastomers and chemicals for the adsorption and permeation of the chemicals by the elastomers. This attack results in swelling and softening, with a reduction of their tensile strength [3].

The temperature and concentration of the chemical medium determine the degree of deterioration. The absorption and desorption of a gas during the change of gas compositions may result in permanent damage of the elastomers.

Some elastomers can be degraded in outdoor conditions when they are exposed to sunlight, ozone, and oxygen. This type of degradation can cause surface cracking, discoloration, significant loss of tensile strength, elongation, and other rubbery properties.

Pure mechanical damage is not a frequent failure mode of elastomers. Most mechanical damage occurs as a result of chemical deterioration of the elastomer. When the elastomer is chemically deteriorated, it is more susceptible to mechanical damage. Elastomers become brittle when cooled below their glass transition temperature, and this can lead to brittle fracture of the elastomers [3].

2.3 LEAKAGE OF STEEL SYSTEMS

In steel and ductile iron systems, leakage mainly occurs through the threads or through mechanical joints. Leakage measurements for steel and ductile iron gas distribution systems suggest that the volume leakage rate for hydrogen is about a factor of 3 higher than that for natural gas [3].

Almost all pipelines are made from ferrous steel, which is sensitive to corrosion in harsh working environments. This is particularly true in the presence of salty water and sulfur-containing media [4].

The mostly used solution for a damaged steel pipe is to remove the pipe entirely or at least the damaged section and then replace it by a new pipe. Welding or fixing the steel patch is difficult especially if the location is underground or underwater.

For this reason, attempts have been made to find effective and safe repair solutions which are light, fast, and easy. For example, fiber-reinforced polymer-based composites can be effectively used for repairing steel pipelines.

Table 2.2 Advantages and disadvantages of pipeline repair resins

Epoxy resin	Polyester	Vinylester
Advantages		
Superior physical and mechanical properties, low shrinkage on curing, good adhesion, wide range of adaptively, better compatibility with carbon fibers, good moisture and chemical resistance	Low cost, available and easily applicable	Better strain and strength performance than polyester, also low cost
Disadvantages		
High costs, may possess corrosive contents and may degrade under ultraviolet light	Moderate strength, low durability, high cure shrinkage, low strain before failure and less compatible with carbon fibers	High shrinkage and exothermic reaction during curing, may require postcuring, lower strain and carbon compatibility than epoxy resins

Source: Shamsuddoha M, Islam MM, Aravinthan T, Manalo A, tak Lau K. Effectiveness of using fibre-reinforced polymer composites for underwater steel pipeline repairs. Compos Struct 2013;100:40–54. http://www.sciencedirect.com/science/article/pii/S0263822312006459, http://dx.doi.org/10.1016/j.compstruct.2012.12.019.

A comprehensive review of the use of fiber-reinforced polymer composites for in-air, underground, and underwater pipeline repairs has been presented [4]. Also, intended future developments have been discussed.

Special composites for pipeline repair are vinylester resins or epoxy resins that are reinforced with glass fibers or carbon fibers. The advantages and disadvantages of the various resins are summarized in Table 2.2.

2.4 LEAK DETECTION TECHNOLOGY

Technologies for the detection of leaks in pipelines have been reviewed [5]. The importance of pipeline leak detection in oil and gas transportation arises from environmental reasons, and also has economic consequences. Also, legislative issues have resulted in a considerable impetus for the adoption of an enhanced monitoring technology.

2.5 FORMATION OF COATINGS

Unsaturated hydrocarbons, such as ethene and propene, can readily polymerize on many surfaces and form impermeable coatings. Such components are present in low concentrations in natural gas [3]. In some pipeline practices, such as propane peak shaving, they can account for more than 1% of the total gas composition.

REFERENCES

[1] Doro-on AM. Risk assessment and security for pipelines, tunnels, and underground rail and transit operations. Boca Raton, FL: CRC Press, Taylor & Francis Group; 2014. ISBN 9781466569348

[2] Zhang S, Zhou W. An efficient methodology for the reliability analysis of corroding pipelines. J Pressure Vessel Tech 2014;136(4):1–7. http://dx.doi.org/10.1115/1.4026797.

[3] Melaina MW, Antonia O, Penev M. Blending hydrogen into natural gas pipeline networks: a review of key issues. Review of Key Issues NREL/TP-5600-51995, National Renewable Energy Laboratory, Golden, CO; 2013. https://www.eere.energy.gov/hydrogenandfuelcells/pdfs/blending_h2_nat_gas_pipeline.pdf.

[4] Shamsuddoha M, Islam MM, Aravinthan T, Manalo A, tak Lau K. Effectiveness of using fibre-reinforced polymer composites for underwater steel pipeline repairs. Compos Struct 2013;100:40–54. http://www.sciencedirect.com/science/article/pii/S0263822312006459, http://dx.doi.org/10.1016/j.compstruct.2012.12.019.

[5] Black P. A review of pipeline leak detection technology. In: Coulbeck B, Evans E, editors. Pipeline systems, Fluid mechanics and its applications, vol. 7. Amsterdam, The Netherlands: Springer; 1992. p. 287–98. ISBN 978-90-481-4143-2, http://dx.doi.org/10.1007/978-94-017-2677-1_23.

CHAPTER 3

Pretreatment Methods

Certain requirements concerning the purity of the product to be transported must be fulfilled. For example, oxygen and sulfur must be controlled effectively. Some additives are presented in Table 3.1.

3.1 GASES

3.1.1 Water Content

For natural gas, the water content should be kept below a certain level (c.f. Section 4.2. Gas Hydrate Formation) to reduce hydrate formation.

3.1.2 Demulsifiers

A gas containing entrained asphaltene-containing hydrocarbons is conditioned for pipeline transportation by injection of a surface-active compound that is composed of a mixture of demulsifying agent, antifoaming agent, dispersant, aromatic solvent, and alcohol solvent, whereby the formation of an emulsion from asphaltene-containing hydrocarbons is prevented [1]. A surface-active composition is shown in Table 3.2.

3.2 HEAVY CRUDE OILS

3.2.1 Emulsions for Heavy Crude Oils

Oil-in-water emulsions provide a cost-effective alternative to the conventional methods—namely, heating or dilution. A typical transport emulsion is composed of 70% crude oil, 30% aqueous phase, and 500–2000 ppm of a stabilizing surfactant formulation [2].

Nonionic surfactants are relatively insensitive to the salt content of the aqueous phase. Ethoxylated alkyl phenols have been used successfully for the formation of stable emulsions that resist inversion.

Guide to the Practical Use of Chemicals in Refineries and Pipelines
http://dx.doi.org/10.1016/B978-0-12-805412-3.00003-9

Table 3.1 Main classes of additives and chemicals used for transport

Additive	Remarks
Drag reducers	For both liquid and gas transport
Pour point depressants	Reduce pour points of waxy crude oils
Odorizing additives	For safety
Gas hydrate inhibitors	For gas and multiphase transport to prevent the formation of hydrate deposits
Surfactants	For multiphase transport of heavy crude oils
Corrosion inhibitors	Both chemical inhibitors and biocides
Paraffin inhibitors	Prevent paraffin deposition

Table 3.2 Surface-active composition for conditioning a gas containing entrained asphaltenes

Action	Compound	Content (wt %)
Demulsifying agent	Solution of a sulfonic acid, a phenolic resin, and alcohol	5–15
Antifoaming agent	Silicone	1–3
Dispersant	Imidazoline	10–40
Aromatic solvent	Benzene, toluene, xylene, and residues of benzene, toluene, and xylene distillation	10–60
Alcohol solvent	From methanol to hexanol	20–60

3.2.2 Activation of Natural Surfactants

With chemical treatment, the natural surfactants in crude oil can be activated [3]. This method has been shown to be effective for highly viscous crude oil from the Orinoco Belt that has traditionally been transported either by heating or dilution. The precursors to the surfactants are preferably the carboxylic acids that occur in the crude oil.

The activation occurs by addition of an aqueous buffer solution [4]. The buffer additive is either sodium hydroxide in combination with sodium bicarbonate or sodium silicate. Water-soluble amines have also been found to be suitable [5]. Additional stabilizers for the emulsion can be multivalent inorganic salts, such as aluminum nitrate [6], in small quantities of 30 ppm.

3.2.3 Low-Temperature Transportation

In addition to the surfactant, a freezing point depressant can be added for low-temperature transportation. Possible depressants include salts, sugars, and alcohols such as glycerol [8].

3.3 GENERAL ASPECTS OF PRETREATMENT

3.3.1 Sulfur Contamination of Refined Products

If refined products, such as gasoline, diesel, jet fuel, or kerosene, are transported in a pipeline, where otherwise sour hydrocarbon fluids are transported, there may be an undesired enrichment of sulfur in the refined products. This can be avoided if the oxygen level of the transported material is maintained below 20 ppm [7].

The dissolved oxygen level in the hydrocarbon product is controlled by reducing the amount of air injection used in mercaptan or disulfide reduction or by the use of oxygen scavengers before the introduction of the refined hydrocarbon product into the pipeline.

3.3.2 Corrosion Prevention

Methods used to control presumptive corrosion include deaeration and dehydration. Carbon dioxide and hydrogen sulfide are the main corrosives in pipelines for natural gas, but they are only aggressive in the presence of water.

Therefore sweetening and drying the gas are useful to prevent corrosion. In oil pipelines, water emulsified in crude oil can cause corrosion problems [9]. Emulsified crude oil in separated produced water is also an environmental and disposal problem.

3.3.3 Pour Point Depressants

Some crude oils are so waxy that their transportation by cold pipelines is very difficult, especially in winter. This is because of the high pour points of such crude oils, which adversely affect the transportation process.

The crystallization of waxes at low temperatures causes reduced liquidity of waxy crude oils, which considerably hampers the transportation of crude oils through long-distance pipelines.

From consideration of all the economic aspects, additive treatment, which depresses the pour point and improves the flow characteristics of the crude oil at lower temperatures, was found to be the most suitable method for transporting waxy crude oil [10].

The state of the art of polymeric wax inhibitors and pour point depressants for waxy crude oils has been reviewed [11]. The structural character, functionality, mechanisms, and the factors that influence the efficiency have been summarized. Most polymeric additives contain both a nonpolar and a polar moiety. Via nucleation, adsorption, co-crystallization and

Table 3.3 Pour point depressants

Chemicals	References
Copolymer of acrylic esters with allyl ethers[a]	[12]
Urea and derivatives	[13]
Homopolymer of acrylic esters	[14]
Grafted ethylene vinyl acetate polymer[b]	[15, 16]
Substituted fullerenes[c]	[17]

[a]In amounts of several hundred parts per million.
[b]Grafted with unsaturated dicarboxylic acid derivates.
[c]Fullerene-aniline, fullerene-phenol adducts.

solubilization interactions, polymeric additives alter the morphology and interface of the precipitated wax crystals, thus inhibiting wax deposition, improving flow, and impeding gel formation. Most polymeric additives fall into the categories of crystalline-amorphous copolymers, ethylene vinyl acetate copolymers, comb polymers, and nanohybrids.

Typical pour point depressants are shown in Table 3.3. The pour point can also be reduced by modification of the crude oil itself (eg, by cracking [18]).

REFERENCES

[1] Mendoza AL. Composition for the conditioning of gas containing entrained asphaltenes. US Patent 5 354 505, assigned to Corpoven, S.A. (Caracas, VE); 1994. http://www.freepatentsonline.com/5354505.html.
[2] Rimmer DP, Gregoli AA, Hamshar JA, Yildirim E. Pipeline emulsion transportation for heavy oils. In: Schramm LL, editor, Emulsions: fundamentals and applications in the petroleum industry, ACS Advances in chemistry series, vol. 231, Washington DC: American Chemical Society; 1992. p. 295–312. ISBN 0-8412-2006-9.
[3] Padron A, Castro L, Zamora G. Transportable and stable hydrocarbons in buffer solution dispersions. In: Proceedings Volume, vol. 2. 6th Unitar et al Heavy Crude & Tar Sands Int. Conf. (Houston, 2/12–17/95); 1995. p. 587–96.
[4] Padron A. Stable emulsion of viscous crude hydrocarbon in aqueous buffer solution and method for forming and transporting same. US Patent 5 526 839, assigned to Maraven, SA (Caracas, VE); 1996. http://www.freepatentsonline.com/5526839.html.
[5] Gregory DM, Taylor SE. Process and apparatus for fuelling a marine vessel. US Patent 6 840 290, assigned to BP Oil International Limited (London, GB); 2005. http://www.freepatentsonline.com/6840290.html.
[6] Rodriguez D, Quintero L, Terrer MT, Jimenez E, Silva F, Salazar J. Process for stabilizing a hydrocarbon in water emulsion and resulting emulsion product. US Patent 4 976 745; 1990. http://www.freepatentsonline.com/4976745.html.
[7] Poirier MA, Falkiner RJ. Method for reducing elemental sulfur pick-up by hydrocarbon fluids in a pipeline (law177). US Patent 5 618 408, assigned to Exxon Research and Engineering Company (Florham Park, NJ); 1997. http://www.freepatentsonline.com/5618408.html.

[8] Gregoli AA, Olah AM. Low-temperature pipeline emulsion transportation enhancement. US Patent 5 156 652, assigned to Canadian Occidental Petroleum Ltd. (Calgary, CA); 1992. http://www.freepatentsonline.com/5156652.html.
[9] Bromley MJ, Gaffney SH, Jackson GE. Oilfield emulsion control, techniques and chemicals used to separate oil and water. In: Proceedings Volume, vol. 3. Inst. Corrosion UK Corrosion 91 Conf. (Manchester, England, 10/22–24/91); 1991.
[10] Sunil Kumar MN. Review on polymeric and copolymeric pour point depressants for waxy crude oils and studies on Bombay high crude oil. Quart J Tech Pap (Inst Petrol) 1989;47–71.
[11] Yang F, Zhao Y, Sjöblom J, Li C, Paso KG. Polymeric wax inhibitors and pour point depressants for waxy crude oils: a critical review. J Dispers Sci Technol 2014; 36(2):213–25, http://dx.doi.org/10.1080/01932691.2014.901917.
[12] Krull M, von Halasz SP, Reimann W, Balzer J, Geiss H. Copolymers of ethylenically unsaturated carboxylic acid esters with polyoxyalkylene ethers of lower, unsaturated alcohols as flow-improving agents for paraffin containing oils. US Patent 5 718 821, assigned to Hoechst Aktiengesellschaft (Frankfurt, DE); 1998. http://www.freepatentsonline.com/5718821.html.
[13] Kissel CL. Method for reducing the pour point of an oil and compositions for use therein. US Patent 5 593 955, assigned to Entek Corporation (Brea, CA); 1997. http://www.freepatentsonline.com/5593955.html.
[14] Meyer G, Kessel D, Rahimian I. The effect of pour point depressants of the polyacrylate-type on crude oil (Wirkung von Stockpunkterniedrigern des Polyacrylat-Typs auf Rohöle). Erdöl Kohle-Erdgas-Petrochem 1995;48(3):135–7.
[15] Balzer J, Feustel M, Krull M, Reimann W. Graft polymers, their preparation and use as pour point depressants and flow improvers for crude oils, residual oils and middle distillates. US Patent 5 439 981, assigned to Hoechst Aktiengesellschaft (DE); 1995. http://www.freepatentsonline.com/5439981.html.
[16] Feustel M, Krull M, Oschmann HJ. Additives for improving the cold flow properties and the storage stability of crude oil. US Patent 6 821 933, assigned to Clariant International Ltd. (Muttenz, CH); 2004. http://www.freepatentsonline.com/6821933.html.
[17] Schriver GW, Patil AO, Martella DJ, Lewtas K. Substituted fullerenes as flow improvers. US Patent 5 454 961, assigned to Exxon Research & Engineering Co. (Florham Park, NJ); 1995. http://www.freepatentsonline.com/5454961.html.
[18] Fung GS, DePalm PE, Sharma P. Pour point depression unit using mild thermal cracker. US Patent 6 337 011, assigned to Halliburton Energy Services, Inc. (Houston, TX); 2002. http://www.freepatentsonline.com/6337011.html.

CHAPTER 4

Gas Hydrate Inhibition

The formation of gas hydrates in natural gas transmission lines was initially described by Hammerschmidt [1] from Texoma Natural Gas Company in 1934. Gas hydrates themselves were discovered by Humphry Davy in 1810. There are monographs on gas hydrates [2].

In the crude oil and natural gas industry, great significance is attached, in particular, to the gas hydrates which form from water and the natural gas constituents methane, ethane, propane, isobutane, *n*-butane, nitrogen, carbon dioxide, and hydrogen sulfide. Especially in modern natural gas extraction, the existence of these gas hydrates constitutes a great problem, especially when wet gas or multiphasic mixtures of water, gas, and alkane mixtures are subjected to low temperatures under high pressure.

As a consequence of the insolubility and crystalline structure of gas hydrates, their formation leads to the blockage of a wide variety of extraction equipment, such as pipelines, valves, and production equipment, in which wet gas or multiphasic mixtures are transported over long distances, as occurs especially in colder regions of Earth or on the seabed [3].

Gas hydrate crystals, when allowed to form and grow inside a conduit such as a pipeline, tend to block or even damage the conduit. To prevent such blockage, the following precautions can be taken [4]:

- removal of free water,
- maintenance of elevated temperatures or reduced pressures,
- addition of melting point depressants.

Furthermore, condensates, hydrates, or ice may form in the course of decompression of natural gas stored in natural reservoirs (eg, in salt caverns). The operation of oil and gas pipelines in the deep sea is significantly complicated by the formation of gas hydrates [5].

4.1 CHARACTERIZATION METHODS

Antiagglomeration testing can be performed with the rocking cell procedure [6]. A rocking cell apparatus been described in the literature [2].

Guide to the Practical Use of Chemicals in Refineries and Pipelines
http://dx.doi.org/10.1016/B978-0-12-805412-3.00004-0

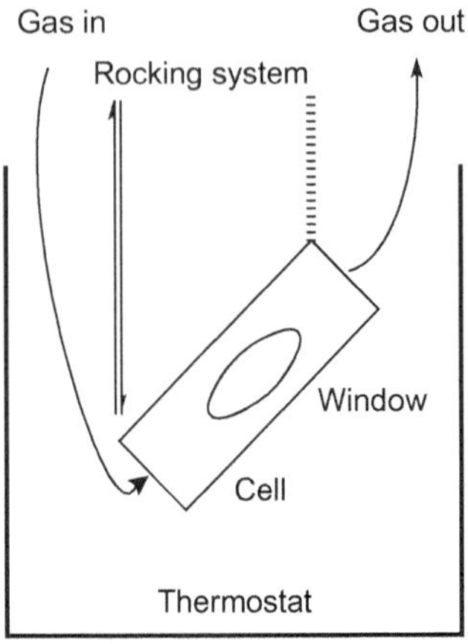

Figure 4.1 Rocking cell system. *(Source: Deaton WM, Frost EMJ. Gas hydrates and their relation to the operation of natural-gas pipe lines. Technical Report BM-Mon-8, Bureau of Mines, Helium Research Center, Amarillo, TX; 1946. http://www.osti.gov/scitech/biblio/7278165).*

The technique was originally described in 1946 [7]. Such a system is schematically shown in Fig. 4.1.

A rocking cell has two parts: the manifold and the cell body [6]. The manifold is made up of stainless steel fittings that are welded together. It has three stems: an inlet stem is used to charge the cell with gas, an outlet stem is used to release the gas from the cell, and the third stem connects to a transducer, which measures the pressure inside the cell.

The cell body has three layers: the outer layer is a polycarbonate tube, the middle layer is made of stainless steel and is connected to the manifold, and the inner layer contains a high-pressure sapphire tube. This sapphire tube can handle up to 200 bar. A small stainless steel ball is located inside the sapphire tube to induce turbulence and to mix the fluids during the rocking.

The technique for rocking cells is based mainly on the visual observation of the motion of a metal ball. The ball moves back and forth in the cell, which rocks at a steady rate in a mixture of oil, water, and gas under pressure and at a declining temperature. The hydrates formed can be seen through a sapphire window [8].

A rating system for low-dose hydrate inhibitors has been presented [6]. This rating system has a scale from 1 to 5. A rating of 1 or 2 signifies failure, a rating of 3 signifies the test has been marginally passed and ratings above 3 signify the test has been passed. In detail, the following characteristics have been given [6]:

Figure 4.2 Sapphire rocking cell. *(Source: PSL Systemtechnik GmbH. Research of gas hydrates with the sapphire rocking cell. Internet, 2014. http://www.psl-systemtechnik.de/sapphirerockingcell0.html?&L=1).*

1. The rolling ball is stuck or the liquid level has dropped below an observable amount.
2. Large to medium-sized agglomerants are present and/or the liquid level has dropped significantly. There is significant resistance to the rolling of the ball in the cell.
3. Medium-sized agglomerants are formed in the viewable area and/or the liquid level has dropped moderately. There is some resistance to the rolling of the ball in the cell.
4. Small agglomerants are formed and/or the liquid level has dropped slightly, but the solution is free flowing without hindrance.
5. There are tiny and well-dispersed hydrates in the hydrocarbon phase, a high liquid level, and free flow of the solution without hindrance.

Meanwhile, highly advanced cell modules are commercially available [9]. For example, the materials are formed from highly corrosion-resistant Hastelloy so that measurements can be conducted in the presence of H_2S. Dual sapphire viewing windows and proximity sensors are included, as well as a center-ported configuration. Another commercially available system is shown in Fig. 4.2.

The use and special aspects of rocking cells under high-pressure conditions have been described [8]. Also, different methods of temperature control have been presented in detail. The standard parameters for testing the formation of gas hydrates are shown in Table 4.1.

Table 4.1 Standard parameters

Parameter	Value
Cell volume	20 mL
Rocking rate	40 rocks per minute
Rocking angle	40 degrees
Rocking ball	Steel ball
Concentration	2500 and 5000 ppm

Source: Lone A. Establishing a new high pressure steel multi-cell rocker rig for kinetic hydrate inhibitor testing; 2011. http://brage.bibsys.no/xmlui/bitstream/handle/11250/182480/Master%20theises%20-%20Astrid%20Lone.pdf?sequence=6.

In the experimental studies, some parameters were changed to see how they would impact gas hydrate formation.

4.2 GAS HYDRATE FORMATION

4.2.1 Simulation of Formation

A recirculation flow loop was used to simulate the flow of light condensate oils in subsea pipelines under hydrate-forming conditions [10]. The pressure drop across sections of the loop and the flow through the section were used to measure the accumulation of hydrates indirectly. Also, visual observations were made.

Two mixtures were prepared in situ by the mixing of light hydrocarbons, nitrogen, and carbon dioxide with a field-stabilized mixture from the North Sea. The vapor pressures of the mixtures were 2 and 4 MPa at 311 K. Experiments were conducted with both mixtures at about 1.5 MPa above their true vapor pressures with water contents of 0.5–5% by volume from 276.5 to 278 K.

The experiments showed no appreciable deposition at a water content of 0.5%. However, at water contents of 3 and 5%, hydrate deposition was observed. An effect of the flow velocity was not found [10].

4.2.2 Risks of Gas Hydrates

Water condensation and hydrate formation at the top of pipelines are serious problems in pipelines with a stratified flow [11]. Tests have been conducted for the assessment of gas hydrate risks in various operational scenarios.

Thermodynamic modeling shows that the presence of a temperature gradient may result in water condensation and also hydrate formation.

A feasibility test on hydrate formation caused by a temperature gradient in a pipeline or a dead leg was developed. A dead leg is the inactive part of the pipeline with no flow or only very low flow. Dead legs are commonly connected to the main pipeline, which has hot reservoir fluid flowing inside it. Because there is no flow or low flow inside dead legs, a temperature gradient may occur therein. This temperature gradient can evaporate the water from the hot reservoir fluid flowing in the main pipeline. This may condense in the cold, top section of the dead leg.

The results provide a better evaluation of the risks involved in various systems and provide guidelines for avoiding the associated problems [11].

4.2.3 Synergism with Corrosion Inhibitors

The possibility of corrosion inhibitors promoting hydrate formation along deepwater pipelines is a flow assurance problem [12].

Several inhibitors were investigated at different concentrations and pressures in a cryogenic sapphire cell at static conditions. All the inhibitors examined had a foamy behavior, but dodecylpyridinium chloride was the compound with the utmost foamy structure.

Changes in the formation temperature established that all the inhibitors promote hydrate formation. However, this occurs at different rates, and also the hydrate formation patterns are different.

The ability to promote hydrate formation can be attributed to the hydrogen bonding properties of corrosion inhibitors, which are required for hydrate formation. The difference in the promotion rate is explained by their different sizes and structures, active functional groups, and affinity for water molecules, which determine the type of hydrogen bonding.

The structure and size of each inhibitor also contribute to its electronegativity and ionization energy. Namely, in some inhibitors there is direct exposure to the nucleus. In other types, the active electrons in the outermost shell are shielded from the direct influence of the attractive force.

4.3 HYDRATE CONTROL

The usual practice for avoiding the plugging of production facilities by hydrates is to add thermodynamic inhibitors, such as methanol or glycol. A newer concept is the injection of low-dose additives: either kinetic inhibitors, which delay nucleation or prevent the growth of hydrate crystals, or hydrate dispersants, which prevent the agglomeration of hydrate particles and allow them to be transported within the flow [13, 14].

Classes of hydrate control agents are shown in Table 4.2, and additives are shown in Table 4.3 and Fig. 4.3.

Table 4.2 Classes of additives for hydrate control

Class	Remark
Thermodynamic inhibitors	Methanol or glycol modifies the stability range of hydrates
Antinucleants	Prevent nucleation of hydrate crystals
Growth modifiers	Control the growth of hydrate crystals
Slurry additives	Limit the droplet size available for hydrate formation
Antiagglomerates	Dispersants that remove hydrates

Source: Goodwin S. Prediction, modelling and management of hydrates using low dosage additives: Pt.1: Additive types and operational implications. In: Proceedings Volume. IBC Tech. Serv. et al Advances in Multiphase Oper. Offshore Conf. (London, England, 11/29–30/95), 1995.

Table 4.3 Additives for hydrate control for pipelines

Additive	Reference
Methanol, glycol[a]	
Poly(*N*-vinyl-2-pyrrolidone)[b]	[15]
Alkyl glycosides[c]	[16]
L-Tyrosine and the methyl ester of L-tyrosine[d]	[17]

[a]Thermodynamic inhibitor.
[b]Growth inhibitor.
[c]C_8–C_{18} alkyl glycoside with glucose, fructose, etc. For example, 1-*O*-octyl-β-D-glucopyranoside or dodecyl-β-maltoside.
[d]Amino acid.

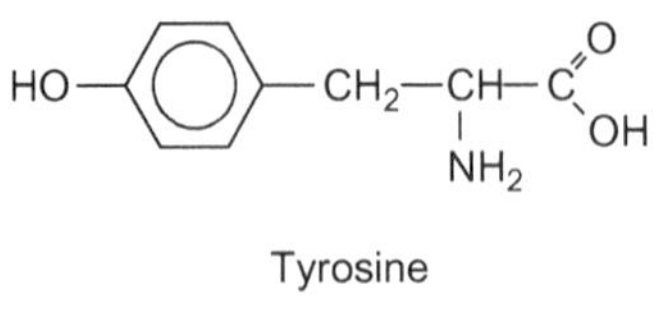

Figure 4.3 Tyrosine.

4.3.1 Thermodynamic Inhibitors

To prevent blocking problems in pipelines, gas hydrate formation in gas pipelines, in the course of transport of multiphasic mixtures or in drilling fluids, can be suppressed by use of relatively large amounts of lower alcohols such as methanol, glycol, or diethylene glycol.

The addition of these additives has the effect that the thermodynamic limit of gas hydrate formation is shifted to lower temperatures and higher pressures (thermodynamic inhibition). However, the addition of these thermodynamic inhibitors causes serious safety problems because of the flash point and the toxicity of the alcohols, logistical problems (large storage tanks, recycling of these solvents), and accordingly, high costs, especially in offshore extraction.

Attempts are therefore being made to replace thermodynamic inhibitors by the addition of additives in amounts of less than 2% in temperature and pressure ranges in which gas hydrates can form. These additives either delay gas hydrate formation (kinetic inhibitors) or keep the gas hydrate agglomerates small and therefore pumpable, so that they can be transported through the pipeline. The inhibitors used either prevent nucleation and the growth of the gas hydrate particles, or they modify the hydrate growth in such a way that relatively small hydrate particles result.

For gas hydrate inhibitors beyond thermodynamic inhibitors there are a multitude of monomeric and also polymeric substance classes, which are kinetic inhibitors or agglomeration inhibitors. Of particular significance in this context are polymers having a carbon backbone which contains both cyclic pyrrolidone or caprolactam moieties and acyclic amide structures in the side groups [3].

The main problem of hydrate formation arises in pipelines transporting natural gas, because gas hydrates are solids and will leave deposits. The solid deposits reduce the effective diameter of the pipeline and can therefore restrict the flow or even clog the pipeline.

Experience indicates that large gas hydrate plugs in gas and oil pipelines form most actively during the period of an unforeseen long shutdown. In static conditions, three types of hydrate crystals can be formed [18]:

1. surface-contact films and massive hydrates, which form by sorption of gas and water molecules on the surfaces of growing crystals,
2. bulk diffusional whisker-like hydrate, which forms both in the volume of gas and in the bulk of liquid water through sorption of molecules on the growing crystal surface and by tunnel sorption of molecules at the base of the crystal,

3. gel-like soft crystals, which form in the bulk of liquid water at a deficiency of dissolved gas in water.

Under the appropriate pressure and temperature conditions for hydrate formation, there may be simultaneous formation of some crystals and decomposition of other crystals.

Antifreeze agents, such as low molecular weight alcohols and glycols, must be added in substantial amounts to be effective, typically on the order of more than 10% [4]. Of course, a serious disadvantage of such amounts is the price of the additive required. In addition, the recovery is complicated and expensive.

An attractive alternative to antifreeze agents is the use of crystal growth inhibitors [4]. Crystal growth inhibitors interfere with the kinetics of the mode of the growth of gas hydrate crystals. These issues will be discussed in detail below.

Dispersing hydrates into a condensate phase by use of antiagglomerants is another alternative to kinetic or thermodynamic inhibitors to prevent hydrate-plug formation in a gas production pipeline [13, 19]. This method has been taken into consideration for pipeline transport. In laboratory experiments it was shown that several commercial dispersants were successful.

At low water concentrations, dodecyl-2-(2-caprolactamyl) ethanamide was shown to be the best commercial available dispersant [13]. Antiagglomerants have been used in several field applications in deepwater systems, both in subsea wells and in dry tree wells, under both flowing and shut-in conditions [19]. The potential advantages of antiagglomerants over methanol include smaller umbilicals, smaller pumps, smaller storage facilities, and less frequent supply trips.

4.3.2 Kinetic Inhibitors

Kinetic hydrate inhibitors have been identified to prevent these hydrate formations so that the fluids can be pumped out before a catastrophic hydrate formation occurs [20]. The kinetic inhibitors prevent or delay hydrate crystal nucleation and disrupt crystal growth. These kinetic hydrate inhibitors contain moieties similar to the gas molecules previously mentioned. It is suspected that these kinetic inhibitors prevent hydrate crystal growth by becoming incorporated in the growing hydrate crystals, thereby disrupting further hydrate crystal growth. The growing hydrate crystals complete a cage by combining with the partial hydrate-like cages around the kinetic hydrate inhibitor moieties containing gas-like groups. These

Table 4.4 Kinetic hydrate inhibitors

Compound
Poly(*N*-methylacrylamide)
Poly(*N*,*N*-dimethylacrylamide)
Poly(*N*-ethylacrylamide)
Poly(*N*,*N*-diethylacrylamide)
Poly(*N*-methyl-*N*-vinylacetamide)
Poly(2-ethyloxazoline)
Poly(*N*-vinylpyrrolidone)
Poly(*N*-vinylcaprolactam)

Source: Panchalingam V, Rudel MG, Bodnar SH. Methods for inhibiting hydrate blockage in oil and gas pipelines using amide compounds. 2008. http://www.freepatentsonline.com/7381689.html.

inhibitors are effective with or without the presence of a liquid hydrocarbon phase, but they are typically less effective in preventing hydrate formation as the production pressure increases.

Examples of polymeric kinetic hydrate inhibitors are shown in Table 4.4. The corresponding monomers are shown in Fig. 4.4.

4.3.3 Antiagglomerate Hydrate Inhibitors

Unlike the kinetic hydrate inhibitors, antiagglomerate hydrate inhibitors are effective only in the presence of an oil phase [20]. These inhibitors do not inhibit the formation of gas hydrates to the same level as kinetic inhibitors; rather their primary activity is in preventing the agglomeration of hydrate crystals. The oil phase provides a transport medium for the hydrates, which are referred to as hydrate slurries, so that the overall viscosity of the medium is kept low and the medium can be transported along the pipeline. As such, the hydrate crystals formed in the water droplets are prevented from agglomerating into a larger crystalline mass.

Antiagglomerate compounds are selected from the group consisting of anionic, cationic, nonionic, and zwitterionic compounds [21]. Antiagglomerate compounds that have been tested as hydrate inhibitors are shown in Table 4.5. Some of these quaternary nitrogen compounds are shown in Fig. 4.5. The effectiveness of hydrate inhibition of some compounds tested is shown in Table 4.6.

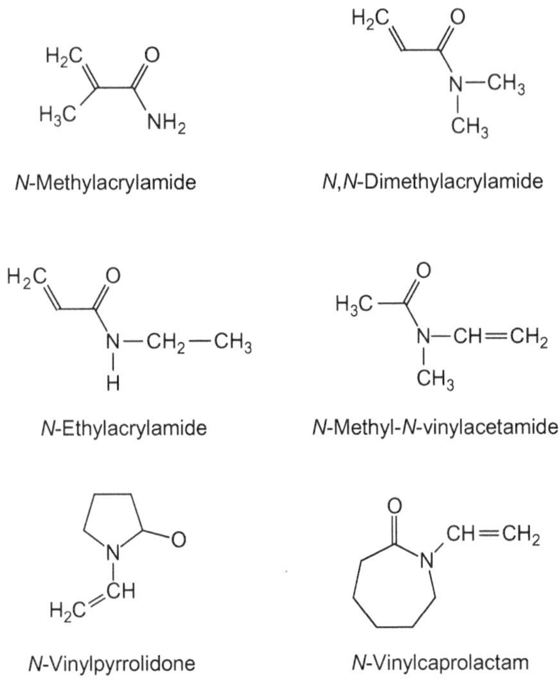

Figure 4.4 Monomers for kinetic hydrate inhibitors.

Other types of antiagglomeration agents are piperazine derivates [6]. The synthesis is shown in Fig. 4.6. These types of inhibitors also act as corrosion inhibitors [22].

4.3.4 Inhibitors with Improved Biodegradability

Common inhibitors such as poly(vinylpyrrolidone) and poly(vinylcaprolactam) are only moderately biodegradable, but nowadays improved biodegradability is desirable [23].

Recently it was found that both water-soluble and oil-soluble pyroglutamic esters and also 1-alkyl-5-oxopyrrolidine-3-carboxylic acid esters are suitable as gas hydrate inhibitors. According to the structure, these esters can delay both the nucleation and the growth of gas hydrates. Further, these compounds suppress the agglomeration of gas hydrates are thus act as antiagglomerants. In addition, these compounds have a significantly improved biodegradability [23, 24].

Nitrogen-containing acids are shown in Fig. 4.7.

Table 4.5 Antiagglomerate compounds

Compound	
Tetrapentylammonium bromide	Tripentylbutylammonium bromide
Triisopentybutylammonium bromide	Triisopentylammonium sulfate
Tetrabutylphosphonium chloride	Tributyldecylammonium bromide
Tributylisopentylammonium bromide	Tripentylammonium sulfate
Tributyltetradecylammonium bromide	Tributylpentylammonium bromide
Tributyltetradecylammonium bromide	Tetrabutylammonium bromide
Tetrabutylammonium chloride	Tributyltetradecylammonium bromide
Tributylhexadecylphosphonium bromide	Tetrabutylammonium toluene-4-sulfonate
Tributylammonium sulfate	Trihexylbutylammonium bromide
Dibutypentylethanolammonium bromide	Tributylheptylammonium bromide
Tetrahexylammonium benzonate	Tetrahexylammonium bromide
Tributylmethylammonium bromide	Dibutyldodecylethylammonium bromide
Tetrahexylammonium chloride	Triisobutylpentylammonium bromide
Triphenylphosphonium bromide	Dipentylammonium sulfate
Tetramethylammonium bromide	Methyltriphenylphosphonium bromide
Tetradecyltrimethylammonium bromide	Butyltriphenylphosphonium bromide
Tetrapropylammonium bromide	Propyltriphenylphosphonium bromide
Tetraphenylphosphonium bromide	Tetraethylammonium bromide
Dodecyltrimethylammonium bromide	2-Dimethylaminoethyltriphenylphos-phonium bromide
Ethyltriphenylphosphonium bromide	Ethylhexadecyldimethylammonium bromide

Source: Klomp UC, Kruka VR, Reijnhart R, Weisenborn AJ. Method for inhibiting the plugging of conduits by gas hydrates. US Patent 5 460 728, assigned to Shell Oil Company (Houston, TX); 1995. http://www.freepatentsonline.com/5460728.html.

Suitable alcohols for the preparation of esters include isobutyl alcohol, 2-ethylhexanol, glycerol, pentaerythritol, dipentaerythritol, and polyglycerol. The esters are prepared in a conventional way from the corresponding alcohols and acids with use of *p*-toluenesulfonic acid as an esterification catalyst.

For the investigation of the kinetic inhibition of the esters, an autoclave was filled with distilled water and gas in a volume ratio of 20:80. For investigations of the agglomerate inhibition, a condensate was additionally added. Finally, natural gas at 50 bar was injected [23].

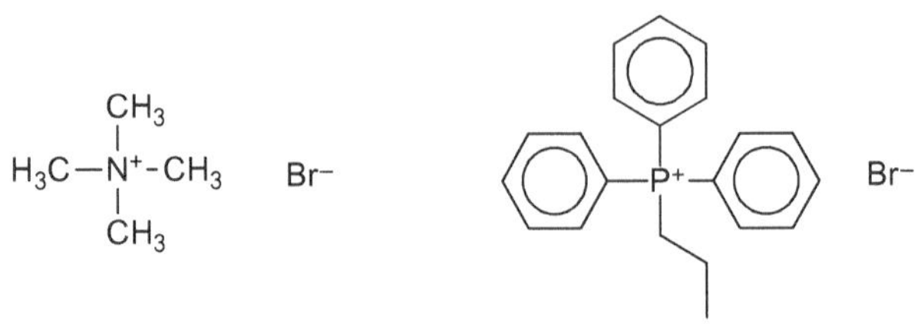

Tetramethyl ammoniumbromide Propyltriphenyl phosphonium bromide

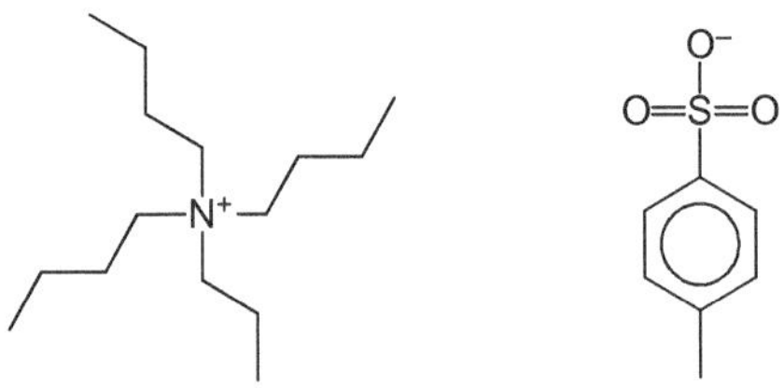

Tetrabutyl ammonium-toluene-4-sulfonate

Figure 4.5 Antiagglomerate compounds.

Table 4.6 Effectiveness of hydrate inhibition

Compound	
Trimethylhexadecylammonium bromide	Partial hydrate blockage at 6°C, later the hydrate broke lose
Dimethylethylhexadecylammonium bromide	Partial hydrate blockage at 4.4°C, later the hydrate broke lose
Dimethylbutylhexadecylammonium bromide	No hydrate blockage at 4.4°C in 18 h
Dimethylbutyloctadecylammonium bromide	No hydrate blockage at 4.4°C in 13 h
Dipropylbutylhexadecylammonium bromide	No hydrate blockage at 4.4°C in 13 h
Dimethyldihexadecylammonium bromide	Complete hydrate blockage at 11°C

Source: Klomp UC, Kruka VR, Reijnhart R, Weisenborn AJ. Method for inhibiting the plugging of conduits by gas hydrates. US Patent 5 460 728, assigned to Shell Oil Company (Houston, TX); 1995. http://www.freepatentsonline.com/5460728.html.

The starting temperature was 20°C. From this temperature, the autoclave was cooled to 4°C within 3 h, then the waser was stirred at 4°C for 18 h and heated back to 20°C within 2 h. When the formation of gas hydrate nuclei occurs during the cooling time, the pressure decreases more than that due to the thermal compression of the gas as such, and a rise in

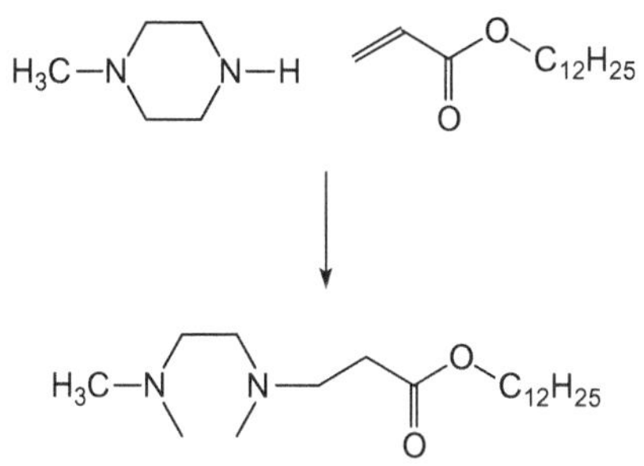

Figure 4.6 Antiagglomeration agents. *(Source: Acosta EJ. Compositions and methods for inhibiting the agglomeration of hydrates in a process. US Patent 8 334 240, assigned to Nalco Company (Naperville, IL); 2012. http://www.freepatentsonline.com/8334240.html).*

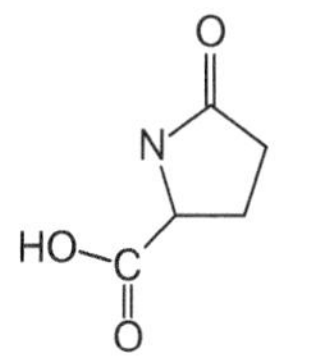

Pyroglutamic acid

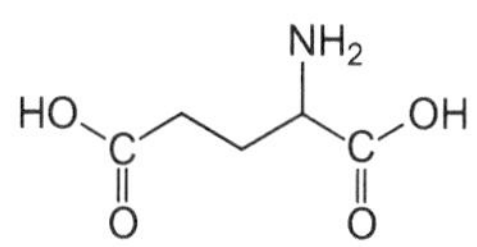

Glutamic acid

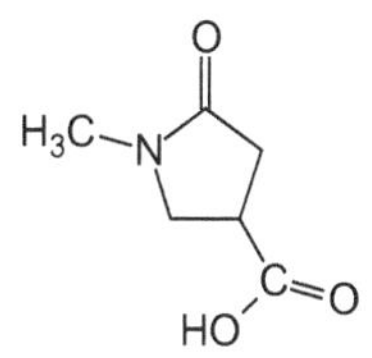

1-Methyl-5-oxopyrrolidine-3-carboxylic acid

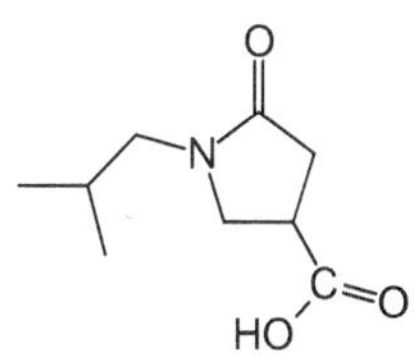

1-Isobutyl-5-oxopyrrolidine-3-carboxylic acid

Figure 4.7 Nitrogen-containing acids.

Table 4.7 Biodegradability (28 days)

Compound	Biodegradability
Poly(vinylpyrrolidone) (comparative compound)	5
Isobutyl 1-methyl-5-oxopyrrolidine-3-carboxylate	81
2-Ethylhexyl 1-methyl-5-oxopyrrolidine-3-carboxylate	66
Glyceryl tri(1-methyl-5-oxopyrrolidine-3-carboxylate)	90
Isobutyl 1-isobutyl-5-oxopyrrolidine-3-carboxylate	73
2-Ethylhexyl 1-isobutyl-5-oxopyrrolidine-3-carboxylate	60
Glyceryl tri(1-isobutyl-5-oxopyrrolidine-3-carboxylate)	85
(Oleyl alcohol+5 EO) 1-isobutyl-5-oxopyrrolidine-3-carboxylate	65
Isobutyl 1-oleyl-5-oxopyrrolidine-3-carboxylate	69
Isobutyl pyroglutamate	65
2-Ethylhexyl pyroglutamate	48
Glyceryl tri(pyroglutamate)	62
Pentaerythrityl di(pyroglutamate)	70
Dipentaerythrityl tetra(pyroglutamate)	71
Polyglycerol ($n = 20$) poly(pyroglutamate)	35
Polyglycerol ($n = 40$) poly(pyroglutamate)	24

5 EO, five ethylene oxide units.
Source: Leinweber D, Roesch A, Feustel M. Use of 1-alkyl-5-oxopyrrolidine-3- carboxylic esters as gas hydrate inhibitors with improved biodegradability. US Patent 8 722 589, assigned to Clariant Finance (BVI) Limited (Tortola, VG); 2014. http://www.freepatentsonline.com/8722589.html.
Source: Leinweber D, Feustel M. Use of pyroglutamic acid esters as gas hydrate inhibitors with improved biodegradability. US Patent 8 735 332, assigned to Clariant Produkte (Deutschland) GmbH (Frankfurt am Main, DE); 2014. http://www.freepatentsonline.com/8735332.html.

the torque is measured together with a slight increase of the temperature. Without inhibitors, further growth and increasing agglomeration of the hydrate nuclei lead rapidly to a further rise in the stirring torque. Eventually, when the mixture is heated, the gas hydrates decompose, so the starting state of the experimental series is attained.

It has been found that all esters mentioned above (from isobutyl alcohol, 2-ethylhexanol, glycerol, pentaerythritol, dipentaerythritol, and polyglycerol.) show performance superior to that of poly(vinylpyrrolidone). The results of studies on biodegradability according to OECD Test No. 306 [25] are shown in Table 4.7.

4.3.5 Polyesters

Special polyesters are active gas hydrate inhibitors. These compounds are synthesized by the condensation of substituted dicarboxylic acids with diols or polyols. Eventually, the free hydroxyl groups are esterified with nitrogen-containing carboxylic acids [26]. Preferred raw materials for these polyesters are summarized in Table 4.8 and some less common compounds are shown in Fig. 4.8.

Table 4.8 Monomers for polyesters

Diacid	Alcohol	*N*-Acid
Tartaric acid	Glycerol	Pyroglutamic acid
Malic acid	Diglycerol	Glutamic acid
Maleic acid	Triglycerol	Aceturic acid
Fumaric acid	Polyglycerol	Aspartic acid
Malonic acid	Trimethylolpropane	
Adipic acid	Pentaerythritol	
Succinic acid	Sorbitol	
Itaconic acid	Ethylene glycol	
Phthalic acid	Diethylene glycol	
	Propylene glycol	
	Dipropylene glycol	
	Butanediol	

Source: Leinweber D, Feustel M. Polymers and their production and use as gas hydrate inhibitors. US Patent 7 893 009, assigned to Clariant Produkte (Deutschland) GmbH (Frankfurt, DE); 2011. http://www.freepatentsonline.com/7893009.html.

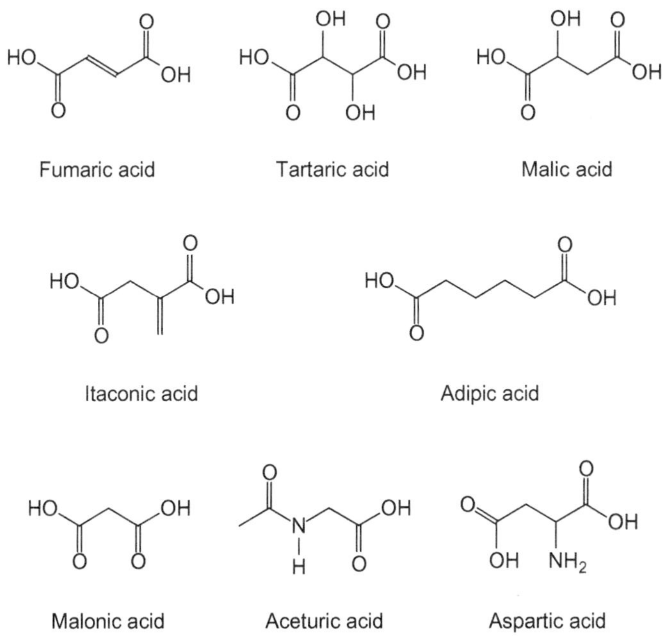

Figure 4.8 Monomers for polyesters.

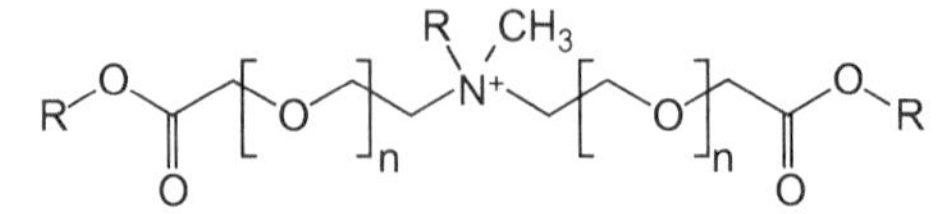

Figure 4.9 Structures of quaternary ammonium hydrate inhibitors. *(Source: Dahlmann U, Feustel M. Additives for inhibiting the formation of gas hydrates. US Patent 8 034 748, assigned to Clariant Produkte (Deutschland) GmbH (Frankfurt, DE); 2011. http://www.freepatentsonline.com/8034748.html).*

4.3.6 Quaternary Ammonium Compounds

Quaternary ammonium compounds for use as hydrate inhibitors can be prepared by the reaction of alkoxylated alkylamines or alkylaminoalkylenamines with monochlorocarboxylic acids to give the corresponding ether carboxylic acids, which are subsequently esterified with alkanols.

Alternatively, bisalkoxylated monoalkylamines can be reacted directly with carboxylic acids or anhydrides and carbonyl chlorides to give the desired esters. Afterward a quaternization with alkylating agents has to be done. The general structure of these compounds is shown in Fig. 4.9.

Several examples of synthesis have been described [27]. In particular, caprylamine, cyclohexylamine, cyclopentylamine, isobutylamine, laurylpropylenediamine, *n*-butylamine, and tallow fat propylenediamine have been used as basic amines.

For use as gas hydrate inhibitors, the quaternary ammonium compounds are preferably dissolved in alcoholic solvents, such methanol, ethanol, propanol, butanol, butyl glycol, isobutyl glycol, butyl diglycol and polyglycols. Further, it has been found that some of these compounds are active as corrosion inhibitors. Therefore, in some cases, the addition of corrosion inhibitors is not necessary.

4.4 TRANSPORTING HYDRATES IN SUSPENSION

Hydrate formation is notorious in the gas and oil industry. The classic method as described is to add products which are thermodynamic hydrate inhibitors (eg, alcohols such as methanol or glycols). However, this solution is very expensive as the quantity of inhibitors to be added may reach 10–40% of the water content. Further, such alcohols pollute the effluents as such types of inhibitors are difficult to recover.

Table 4.9 Alcohols and acids

Acids	Alcohols
Octanoic acid	Ethylene glycol
Undecenoic acid	Propylene glycol
Lauric acid	Neopentyl glycol
Oleic acid	Glycerol
Ricinoleic acid	Trimethylolpropane
Arachidic acid	Pentaerythritol
Gadoleic acids	Sorbitol
Erucic acid	Polyglycerols

Source: Sinquin A, Dalmazzone C, Audibert A, Pauchard V. Method for transporting hydrates in suspension in production effluents employing a nonpolluting additive. US Patent 7 851 413, assigned to IFP Energies Nouvelles (Rueil Malmaison Cedex, FR); 2010. http://www.freepatentsonline.com/7851413.html.

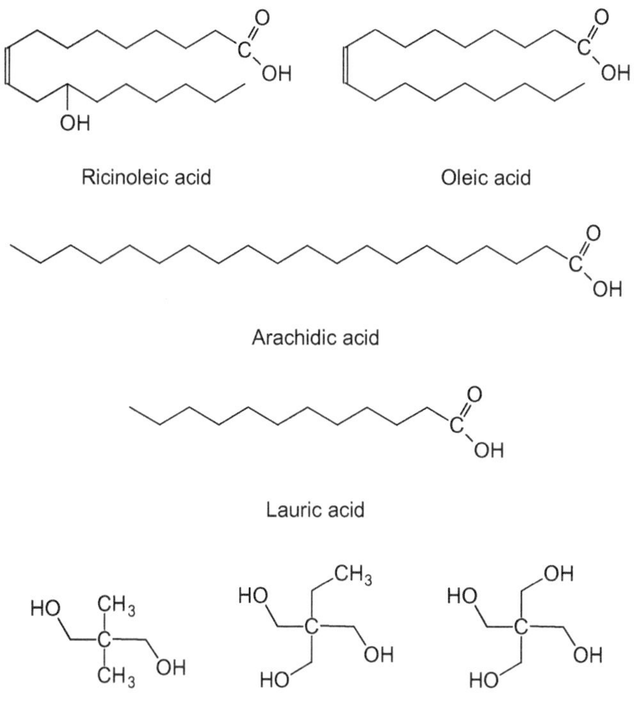

Figure 4.10 Alcohols and acids.

To transport hydrates in suspension in a fluid comprising water, gas, and a liquid hydrocarbon, a nonpolluting composition consisting essentially of a mixture of an ester associated with a nonionic cosurfactant of an oligomeric carboxylic acid type is incorporated into the fluid. The composition may be added in a concentration of only 0.1–5%. Alcohols and acids are shown in Table 4.9 and Fig. 4.10

REFERENCES

[1] Hammerschmidt EG. Formation of gas hydrates in natural gas transmission lines. Indust Eng Chem 1934;26(8):851–5. http://dx.doi.org/10.1021/ie50296a010.

[2] Sloan ED, Koh CA. Clathrate hydrates of natural gases. 3rd ed., no. 119 in Chemical Industries, Boca Raton, FL: CRC Press; 2008. ISBN 9780849390784.

[3] Dahlmann U, Feustel M. Corrosion and gas hydrate inhibitors having improved water solubility and increased biodegradability. US Patent 7 341 617, assigned to Clariant Produkte (Deutschland) GmbH (Sulzbach, DE); 2008. http://www.freepatentsonline.com/7341617.html.

[4] Klomp UC, Kruka VR, Reijnhart R, Weisenborn AJ. Method for inhibiting the plugging of conduits by gas hydrates. US Patent 5 648 575, assigned to Shell Oil Company (Houston, TX); 1997. http://www.freepatentsonline.com/5648575.html.

[5] Mei DH, Liao J, Yang JT, Guo TM, Experimental and modeling studies on the hydrate formation of a methane + nitrogen gas mixture in the presence of aqueous electrolyte solutions. Ind Eng Chem Res 1996;35(11):4342–7.

[6] Acosta EJ. Compositions and methods for inhibiting the agglomeration of hydrates in a process. US Patent 8 334 240, assigned to Nalco Company (Naperville, IL); 2012. http://www.freepatentsonline.com/8334240.html.

[7] Deaton WM, Frost EMJ. Gas hydrates and their relation to the operation of natural-gas pipe lines. Technical Report BM-Mon-8, Bureau of Mines, Helium Research Center, Amarillo, TX; 1946. http://www.osti.gov/scitech/biblio/7278165.

[8] Lone A. Establishing a new high pressure steel multi-cell rocker rig for kinetic hydrate inhibitor testing; 2011. http://brage.bibsys.no/xmlui/bitstream/handle/11250/182480/Master%20theises%20-%20Astrid%20Lone.pdf?sequence=6.

[9] World Houston International Business Center, Rocking cell system for testing low dosage hydrate inhibitors. Oil Gas Petrochemequip 2009;55(11). http://www.ogpe.com/articles/print/volume-55/issue-11/exploration-drilling-production/rocking-cell-system.html.

[10] Dholabhai PD, Kalogerakis N, Bishnoi PR, Evaluation of gas hydrate formation and deposition in condensate pipelines: pilot plant studies. SPE Product Facil 1993;8(03):185–90. http://dx.doi.org/10.2118/22829-PA.

[11] Nazeri M, Tohidi B, Chapoy A, et al. An evaluation of risk of hydrate formation at the top of a pipeline. In: SPE Asia Pacific oil and gas conference and exhibition, Society of Petroleum Engineers; 2012. p. 22–4. http://dx.doi.org/10.2118/160404-PA.

[12] Obanijesu EO, Gubner R, Barifcani A, Pareek V, Tade MO, The influence of corrosion inhibitors on hydrate formation temperature along the subsea natural gas pipelines. J Petrol Sci Eng 2014;120(0):239–52. http://www.sciencedirect.com/science/article/pii/S0920410514001429, http://dx.doi.org/10.1016/j.petrol.2014.05.025.

[13] Huo Z, Freer E, Lamar M, Knauss DM, Sloan Jr ED, Sannigrahi B, Hydrate plug prevention by anti-agglomeration. Chem Eng Sci 2001;56(17):4979–91.

[14] Palermo T, Sinquin A, Dhulesia H, Fourest JM. Pilot loop tests of new additives preventing hydrate plugs formation. In: Proceedings Volume. 8th Bhr Group Ltd et al Multiphase 97 Int. Conf. (Cannes, France, 6/18–20/97); 1997. p. 133–47. ISBN 1-86058-089-0.
[15] Sloan Jr ED. Method for controlling clathrate hydrates in fluid systems. US Patent 5 880 319, assigned to Colorado School of Mines (Golden, CO); 1999. http://www.freepatentsonline.com/5880319.html.
[16] Thieu V, Bakeev KN, Shih JS. Gas hydrate inhibitor. US Patent 6 359 047, assigned to ISP Investments Inc. (Wilmington, DE); 2002. http://www.freepatentsonline.com/6359047.html.
[17] Duncum SN, Edwards AR, Osborne CG. Method for inhibiting hydrate formation. US Patent 5 331 105, assigned to The British Petroleum Company p.l.C. (London, GB2); 1994. http://www.freepatentsonline.com/5331105.html.
[18] Makogon YF. Formation of hydrates in shut-down pipelines in offshore conditions. In: Proceedings Volume, vol. 4. 28th Annu. SPE et al Offshore Technol. Conf. (Houston, 5/6–9/96); 1996. p. 749–56.
[19] Frostman LM, Przybylinski JL. Successful applications of anti-agglomerant hydrate inhibitors. In: Proceedings Volume. SPE Oilfield Chem. Int. Symp. (Houston, TX, 2/13–16/2001); 2001.
[20] Panchalingam V, Rudel MG, Bodnar SH. Methods for inhibiting hydrate blockage in oil and gas pipelines using amide compounds; 2008. http://www.freepatentsonline.com/7381689.html.
[21] Rabeony M, Peiffer DG, Costello CA, Colle KS, Wright PJ, Talley LD. Gas hydrate anti-agglomerates. US Patent 6 015 929, assigned to Exxon Research and Engineering Co. (Florham Park, NJ); 2000. http://www.freepatentsonline.com/6015929.html.
[22] Acosta EJ, Clark JC. Corrosion inhibitors for a fluid. US Patent 8 105 988, assigned to Nalco Company (Naperville, IL); 2012. http://www.freepatentsonline.com/8105988.html.
[23] Leinweber D, Feustel M. Use of pyroglutamic acid esters as gas hydrate inhibitors with improved biodegradability. US Patent 8 735 332, assigned to Clariant Produkte (Deutschland) GmbH (Frankfurt am Main, DE); 2014. http://www.freepatentsonline.com/8735332.html.
[24] Leinweber D, Roesch A, Feustel M. Use of 1-alkyl-5-oxopyrrolidine-3-carboxylic esters as gas hydrate inhibitors with improved biodegradability. US Patent 8 722 589, assigned to Clariant Finance (BVI) Limited (Tortola, VG); 2014. http://www.freepatentsonline.com/8722589.html.
[25] OECD. Biodegradabiltiy in seawater. OECD Guideline for Testing of Chemicals OECD TG 306. Paris: Organisation for Economic Co-operation and Development (OECD); 1992. http://www.oecd.org/chemicalsafety/testing/34898616.pdf.
[26] Leinweber D, Feustel M. Polymers and their production and use as gas hydrate inhibitors. US Patent 7 893 009, assigned to Clariant Produkte (Deutschland) GmbH (Frankfurt, DE); 2011. http://www.freepatentsonline.com/7893009.html.
[27] Dahlmann U, Feustel M. Additives for inhibiting the formation of gas hydrates. US Patent 8 034 748, assigned to Clariant Produkte (Deutschland) GmbH (Frankfurt, DE); 2011. http://www.freepatentsonline.com/8034748.html.

CHAPTER 5

Corrosion in Pipelines

There are monographs and reviews concerning the problems of corrosion in systems for storage and transportation [1–4].

The mechanisms of stress corrosion cracking of pipeline steels have been reviewed [5]. The aspects and the influencing factors for the high pH stress corrosion cracking that arises from a concentrated bicarbonate-carbonate solution, and also for low pH stress corrosion cracking due to a more dilute solution have been exemplified. In the case of high pH stress corrosion cracking, anodic dissolution for crack initiation and propagation occurs. In contrast, low pH stress corrosion cracking is believed to occur by the dissolution of the crack tip and the sides that allows the entry of hydrogen into the pipeline steel.

Hydrogen is considered as an alternative energy source, because the monetary and environmental costs associated with the use of natural gas and petroleum fuels are continuously increasing. Gaseous hydrogen offers an efficient way of storing the energy generated by wind and solar farms through networks of pipelines and caverns. The fatigue crack growth for pipeline steels exposed to hydrogen has been reviewed in detail [6].

Also, the mechanisms of initiation and prevention of top-of-the-line corrosion have been reviewed [7]. Recent multifaceted techniques have been described and exemplified. Corrosion reactions are common to many pipeline conditions and are mostly

$$\begin{aligned} Fe + H_2S &\rightarrow FeS + H_2 \\ Fe + CO_2 + H_2O &\rightarrow FeCO_3 + H_2 \\ 2Fe + O_2 + 4H^+ &\rightarrow 2Fe^{2+} + H_2 + 2H_2O \end{aligned} \quad \text{(i)}$$

Carbon dioxide will dissolve in the water, thus forming carbonic acid, which makes the fluid acidic [3]. The corrosion caused by carbon dioxide is influenced by temperature, increase in pH, the composition of the aqueous stream, the presence of nonaqueous phases, the flow conditions, and metal characteristics [8].

Guide to the Practical Use of Chemicals in Refineries and Pipelines
http://dx.doi.org/10.1016/B978-0-12-805412-3.00005-2

Carbon dioxide corrosion can appear in two principal forms: Pitting is a local attack that results in rapid penetration and removal of metal in a small discrete area. Secondly, mesa attack may occur; this is a form of a local carbon dioxide corrosion during medium-flow conditions [9].

The aspects of pitting corrosion of metals have been reviewed. An overview of the critical factors that influence the pitting corrosion of metals has been presented [10].

5.1 HISTORY

The history of corrosion inhibitors and neutralizers and their invention, development, and application in the petroleum industry have been reviewed by Fisher [11]. Early corrosion inhibitor applications in each of the various segments of the industry, including oil wells, natural gas plants, refineries, and product pipelines, were reviewed.

Corrosion and scale deposition are the two most costly problems in the oil industry. Corrodible surfaces are found throughout production, transport, and refining equipment. The *Corrosion and Scale Handbook* [12] gives an overview of corrosion problems and methods of corrosion prevention.

5.2 TEST METHODS

The issues for selecting a specific method in order to determine the corrosivity have been described [13]. Basically, in the absence of water, the crude oil is noncorrosive. The corrosivity of crude oil containing water can be determined by a combination of three properties [13]:

1. the type of emulsion formed between oil and water,
2. the wettability of the steel surface,
3. the corrosivity of the water phase in the presence of oil.

Laboratory tests for inhibitor evaluation consist of two main components: the laboratory method and the measurement technique [14]. The laboratory methods and measurement techniques for inhibitor evaluation in multiphase systems are shown in Table 5.1.

The wheel test method is used to evaluate film-persistent corrosion inhibitors for oilfield applications [15]; however, this test method is not uniformly accepted throughout the oil industry.

Several laboratory tests for the efficiency of corrosion inhibitors for oilfield pipelines have been described [16–20].

Table 5.1 Laboratory methods and measurement techniques for corrosion inhibitor evaluation

Method	Measurement technique	Phases
Rotating cylinder electrode	Mass loss, electrochemical	Aqueous phase
Rotating cylinder electrode	Mass loss	Aqueous/oil phase
Jet impingement	Mass loss, electrochemical	Aqueous phase
Jet impingement	Mass loss	Aqueous/oil phase
Jet impingement	Electrochemical measurements	Aqueous phase
Rotating cage	Mass loss	Aqueous or aqueous/oil phase

5.3 THERMODYNAMIC AND KINETIC ASPECTS

The corrosion kinetics and thermodynamics of carbon dioxide and hydrogen sulfide have been reported, and the corrosion effects and the prediction methods in oil and gas pipeline systems have been described [21].

The presence of carbon dioxide, hydrogen sulfide, and free water can cause severe corrosion problems in oil and gas pipelines. The internal corrosion in pipelines is influenced by temperature, pH, and the content of the corrosive gases. When an artificial galvanic cell is formed because of the presence of ions in the pipeline, this effects the major amount of corrosion.

The corrosion rate can be substantially reduced when $FeCO_3$, an initial corrosion product, is formed; this can precipitate on the steel surface and eventually form a dense and protective corrosion product film. This occurs more easily at high temperature or high pH when a water phase is present [21].

Thermodynamic principles can explain a corrosion situation in terms of the solubility of chemical species and reactions associated with corrosion processes. However, thermodynamic principles cannot be used to predict corrosion phenomena or the rate of corrosion [22].

There are three basic kinetic laws that characterize the oxidation rates of pure metals:

1. parabolic rate law,
2. logarithmic rate law,
3. linear rate law.

The parabolic rate law assumes that the concentrations of diffusing species at the oxide-metal and oxide-gas interfaces are constant, and that the oxide layer is uniform, continuous, and of the single phase type. This law is applicable to high-temperature engineering problems.

Both the logarithmic rate law and the linear rate law are empirical relationships. Metals with linear oxidation kinetics have a tendency to undergo a so-called catastrophic oxidation at a certain temperature [23].

5.4 MICROBIOLOGICALLY INFLUENCED CORROSION

Microbiologically influenced corrosion is responsible for most of the internal corrosion problems in oil transportation pipelines and storage tanks [24]. One problematic area in treating gas lines is the occurrence of the stratification of water in the line. Under these conditions, corrosion inhibitors do not come into contact properly and oil and inhibitors undergo degradation.

The role of bacteria in oil degradation, the consequences of oil degradation in fuel systems, and its influence on corrosion have been explained in detail. Also, factors influencing the degradation of oil and corrosion inhibitors have been discussed.

Many of the issues of misapplication of biocides or inhibitors occur because the characteristics of biocides and inhibitors are not considered before use in the pipeline industry [24].

Microbiologically influenced corrosion almost always acts in concert with other corrosion mechanisms and may, at times, appear to be crevice corrosion, underdeposit acid attack, oxygen concentration cell corrosion, ion concentration cell corrosion, and CO_2 corrosion [25].

In most cases in which microbiologically influenced corrosion is found on external surfaces, this corrosion is associated with disbonded coatings or other areas that are shielded from the potentially protective action of cathodic protection. Furthermore, pipelines are often in contact with wet clays, which have little scaling potential.

5.4.1 Bacteria

Microbiologically influenced corrosion is a problem commonly encountered in facilities in the oil and gas industries [26].

The biodegradation of naphtha by bacterial isolates was characterized by gas chromatography coupled with mass spectrometry [27].

DNA Sequencing

The bacterial enumeration and identification in diesel and naphtha pipelines located in the northwest and southwest regions of India has been reported. The traditional cultivation technique and 16S ribosomal DNA gene sequencing has been used. DNA sequencing was done with a Genetic Analyzer from PE Applied Biosystems.

The phylogenetic analysis of 16S ribosomal RNA sequences of the isolated species was done. The sequences obtained were analyzed with a BLAST search. The samples obtained from the diesel- and naphtha-transporting pipelines showed the occurrence of 11 bacterial species, as summarized in Table 5.2. Sulfate-reducing bacteria were not detected in samples from pipelines. The dominant bacterial species identified in the samples were *Bacillus cereus* and *Serratia marcescens*.

It has been concluded that several types of bacteria may be involved in the biocorrosion arising from natural biofilms that develop in the pipelines. Further, localized pitting was observed by scanning electron microscopy analysis [26].

5.5 CORROSION INHIBITION METHODS

The rapid expansion of the production of oil and gas from unconventional sources requires the construction of new pipelines. This requires the use of new and innovative corrosion protection methods for pipelines. This is necessary because of the complex corrosive environment of multiphase pipelines and stratified flow pipelines.

Conventional corrosion-inhibition methods, such as continuous or intermittent injection, may not be effective for inhibiting corrosion of the pipelines because physical contact of the inhibitors is difficult to achieve

Table 5.2 Bacterial species in pipelines

Species	
Serratia marcescens ACE2	*Bacillus subtilis* AR12
Bacillus cereus ACE4	*Pseudomonas aeruginosa* AI1
Klebsiella oxytoca ACP	*Pseudomonas stutzeri* AP2
Bacillus litoralis AN1	*Bacillus* sp.
Bacillus pumilus AR2	*Bacillus carboniphilus* AR3
Bacillus megaterium AR4	

Source: Rajasekar A, Anandkumar B, Maruthamuthu S, Ting YP, Rahman P. Characterization of corrosive bacterial consortia isolated from petroleum-product-transporting pipelines. Appl Microbiol Biotechnol 2010;85(4):1175–88. http://dx.doi.org/10.1007/s00253-009-2289-9.

in multiphase flowing systems. Further, the pipelines must be piggable for proper cleaning.

Innovative methods have been developed to provide effective and long-lasting protection to the surface of the pipeline. These are [7]:

- the use of pig trains with inhibitor slugs,
- improved inhibitor chemistries,
- inhibitors added to gels and foams,
- special pigs that provide a 360-degree spray of inhibitor chemicals.

Commercially available methods of corrosion treatment are [7]:

- the injection of contemporary inhibitors using common equipment and systems;
- gels, which exhibit excellent surface contact and can be used with pigs for cleaning;
- spray pigs, which have excellent surface coverage and use less inhibitor for coverage;
- batch treatments with slugs and pigs (these have been used for many years.

Experimental methods of corrosion treatment are [7]:

- foams, which use less liquid and fill all parts of the line;
- injection of advanced inhibitors, which can be used with conventional equipment;
- spray pig with a two-step inhibitor which has excellent surface coverage and a long film life.

5.6 CORROSION CONTROL

Coatings, cathodic protection, and chemical additives are used extensively to prevent internal and external pipeline corrosion. The excessive use of incompatible chemical additives has caused severe problems in gas-transporting systems.

The costs arising from these problems often exceed the costs of the chemicals themselves. The careful evaluation and selection of chemical additives can minimize these problems and reduce operating costs [28].

5.6.1 Crude Oil Treatment

Some crude oils contain organic compounds that are corrosive. In particular, these include naphthenic acids. Such crude oils cause problems in transportation, refining, and processing. The naphthenic acid content can be reduced simply by treatment with alcohol, such as methanol, to form the corresponding ester.

The treatment temperatures will preferably be around 350°C. Pressures of about 100–300 kPa are typical and generally result from the system itself [29].

Synergism With Drag Reducers

Copolymers of acrylamide and acrylic acid that are added as drag reducers also enhance the activity of corrosion inhibitors in brine solutions by decreasing turbulence so that the corrosion inhibitor can more thoroughly contact the internal surface of the pipeline. Therefore the copolymer acts synergistically with corrosion inhibitors to increase their efficiencies [30].

The effect of a chemical drag reducer on oxygen corrosion of carbon steel has been investigated in a large-scale flow loop simulating a seawater injection line. A drag reduction of up to 48% was achieved. In addition, reduction of corrosion occurred [31].

5.6.2 Coatings

Coatings protect the wall material by preventing aggressive substances from coming in contact with the wall. A coating is actually not an additive, but this section is included for completeness. A polyethylene coating with special antioxidant stabilizers has excellent resistance to thermal oxidation at 80°C.

A polypropylene coating is suitable for external corrosion protection of steel pipelines at temperatures between −3 and 120°C. Coatings are summarized in Table 5.3.

Table 5.3 Coatings for pipelines

Coating material	Remarks	Reference
Bitumen	Oxidized bitumen with hexamethylenetetramine novolac resin	[32]
Concrete	Styrene-butadiene copolymer latex addition on centrifugally cast concrete	[33]
Epoxide resins	Glass fiber reinforced	[34]
Polyethylene		[35]
Polypropylene	External corrosion protection	[36]
Polyurethane foams	Insulating systems for high-temperature marine pipelines	[37]
Polyurea		[38]

Alternative Plastic Materials

Fiber-reinforced epoxide pipes have many advantages for the petroleum industry. These pipes offer corrosion resistance, high strength-to-weight ratio, light weight, desirable electrical properties, dimensional stability, pressure and temperature stability in given ranges, and low maintenance costs.

5.6.3 Acid Gas Removal

Acidic compounds such as CO_2, H_2S, and COS in acid gas streams can be separated or removed from the streams by scrubbing with a liquid absorbent medium [39].

The scrubbing process is extensively used in petroleum production operations and in petrochemical refining processes. A number of different technologies are available for removing acid gases such as carbon dioxide, hydrogen sulfide, and carbonyl sulfide. These processes include chemical absorption, physical absorption, cryogenic distillation (ie, the Ryan-Holmes process [40]), and membrane system separation.

Of these, amine separation is a highly developed technology with a number of competing processes in hand using various amine sorbents such as monoethanolamine, diethanolamine, triethanolamine, methyldiethanolamine, diisopropylamine, diglycolamine, 2-amino-2-methyl-1-propanol, and piperazine. Alkanolamine-type absorbents are usually preferred in view of the greater degree of solubility of the reaction products formed in the absorption process in water.

The amine purification process usually contacts the gas mixture countercurrently with an aqueous solution of the amine in an absorber tower. The liquid amine stream is then regenerated by desorption of the absorbed gases in a separate tower, with the regenerated amine and the desorbed gases leaving the tower as separate streams. The various gas purification processes which are available have been described [41].

An operational problem often encountered with acid gas abatement processes is corrosion of carbon steel and other low-alloy steels that are used in the construction of the piping and vessels of the treatment unit. The corrosion can be attributable to one or more of the following [39]:

1. decomposition of the scrubbing agent solvent,
2. reaction of the acidic components of the gas and the scrubbing agent solvent,
3. direct attack by the acidic components in the gases.

Another problem is the accumulation of heat-stable salts formed, for example, by the ingress of reactive contaminants and by degradation of amine, which in itself can lead to a higher corrosion rate and equipment damage. Accumulation of heat-stable salts also consumes scrubbing agent solvents, reducing the amount available for acid gas absorption. Management of the levels of heat-stable salts can be achieved through a reclamation process or replacement of part of the amine inventory with fresh, uncontaminated amine.

However, use of corrosion-inhibition technology has not been a standard practice in amine treating systems because of the high cost of most programs and undesirable side effects (eg, foaming). As such, for a typical acid gas treating facility (eg, an amine treatment facility) the primary mitigation strategy for corrosion of carbon steel or other low-alloy steels has been to replace these materials with more corrosion-resistant stainless steel.

A more recent corrosion control program considered for acid gas treating facilities involves the addition of soluble sodium tetrasulfide to the circulating amine [39]. This program was shown in laboratory and field studies to form protective iron sulfide layers on carbon steel to reduce corrosion rates. Unfortunately, the high cost of the additive makes this concept economically infeasible. As an alternative, the introduction of soluble polysulfide ions was determined to be an effective corrosion mitigation strategy as they are effective in forming protective sulfides on the walls of the equipment.

The addition of ammonium or metal polysulfides of forming polysulfide ions into the circulating amine treating solution has been described [42, 43]. Although theoretically successful, the cost of these various chemical addition techniques has proved to be prohibitive in relation to their benefits and has resulted in limited commercial applicability to date.

Special alkylamine alkyloxy alkyl ethers, which are selective for the sorption of H_2S from acidic gas mixtures containing CO_2, have been described [44]: bis(*tert*-butylamino)ethoxyethane, bis(2-chloroethoxy)ethane, and ethoxyethoxyethanol-*tert*-butylamine.

It has been found that polysulfide ions can be generated in situ in the form of an amine polysulfide by the electrochemical oxidation of the hydrosulfide HS^- or sulfide S^{2-} ions formed from dissociated hydrogen sulfide in the circulating amine treating solution process stream [39]. The electrochemical generation of polysulfide ions may be done by the anodic oxidation of the hydrosulfide ions under the conditions described in the literature [45].

Effective amounts of circulating polysulfide ions can be achieved from the oxidation to obtain effective corrosion control, as well as to manage cyanide and mercury levels, while obviating the need to purchase polysulfides [39].

One achieves the oxidation of the ions by flowing the rich amine solution from the absorption section of the unit through an electrochemical cell to generate the polysulfide ions, which then react with the metal surfaces of the equipment to form a protective polysulfide coating [39].

Cyanide Control

Polysulfide ions can also manage the impact of cyanide ions in the scrubbing agent solvent to control the negative effects of cyanide in acid gas treatment processes. The introduction of elemental sulfur to generate polysulfide ions can prevent hydrogen embrittlement in the upper parts of a regeneration tower. The HCN amine salt partially decomposes in the upper section of the regenerator to give HCN, which produces H^+ ions. H^+ ions penetrate the metal structure and are converted to a larger H_2 molecule, causing a tremendous force inside the metal that causes it to split. The polysulfide ions, however, can be used to convert cyanide ions to the more benign thiocyanate ions and a polysulfide ion with one less sulfur atom [39]:

$$CN^- + S_n^{2-} \rightarrow SCN^- + S_{n-1}^{2-} \qquad \text{(ii)}$$

Mercury Control

The polysulfide ions may also be capable of providing mercury control when this metal, which is found in at least trace quantities in all crude oils and condensates, is present. The polysulfides provide the ability to use existing gas treatment equipment to precipitate soluble or reactive mercury as an insoluble mercury sulfide that can be subsequently removed by filtration. Polysulfide ions can react with elemental mercury to form insoluble mercury sulfide via, for example, the following reaction [39]:

$$Hg + S_n^{2-} \rightarrow HgS + S_{n-1}^{2-} \qquad \text{(iii)}$$

5.7 CLASSIFICATION OF CORROSION INHIBITORS

Inhibitors may be classified according to their solution properties as oil-soluble inhibitors, water-soluble inhibitors, or dispersible inhibitors. Chemical inhibitors act as film formers to protect the surface of the pipeline. Corrosion inhibitors, used for the protection of oil pipelines, are often complex mixtures. Further, corrosion inhibitors have been divided into many groups, such as [46]:

- cathodic and anodic inhibitors,
- inorganic and organic corrosion inhibitors,
- film-forming and non–film-forming inhibitors.

Low molecular weight corrosion inhibitors often change the surface tension of water. These inhibitors are acting as surfactants, as they form a protective layer on the metal surfaces [46]. Polymeric corrosion inhibitors act in the same way as ordinary low molecular weight inhibitors.

Polymeric film-forming corrosion inhibitors are different from polymer coatings as they exhibit a specific interaction with the surface before the dry film is formed. Polymeric corrosion inhibitors may not form a barrier layer against oxygen and water. Instead, they change the corrosion potential of the metal [46]. From the chemists' point of view, corrosion inhibitors can be classified into the following broad groups:

- amides and imidazolines,
- salts of nitrogenous molecules with carboxylic acids (ie, fatty acids and naphthenic acids),
- nitrogen quaternaries
- polyoxylated amines, amides, and imidazolines,
- nitrogen heterocyclics.

A few inhibitors are shown in Fig. 5.1.

Most of inhibitors used in oil production systems are nitrogenous and can be classified as shown in Table 5.4. Typical corrosion inhibitors are shown in Table 5.5 and Fig. 5.2.

Further compounds are summarized in Table 5.6. Some of these compounds are shown in Fig. 5.3.

Acetylenic alcohol can be prepared in a process reactor system as shown in Fig. 5.4. A tank (10) contains a precursor comprising sodium chlorate, sodium chlorite, and optional sodium chloride in a water base. This mixture is pumped by a precursor pump (12) in a precursor feedstream (14) through a check valve (16). Water from a tank (18) is pumped by a water pump (20) in a feedstream (22) and combined with the precursor feedstream (14) to form a further feedstream (24). Another tank (26) contains hydrochloric acid, which is pumped by an acid pump (28) in a feedstream (30). Feedstream (24) is pumped past a check valve (32) and feedstream (30) is pumped past a check valve (34), where they are combined to form a reaction stream (36). The reaction stream (36) passes into a high-pressure reactor (38) that is equipped with an in-line mixer.

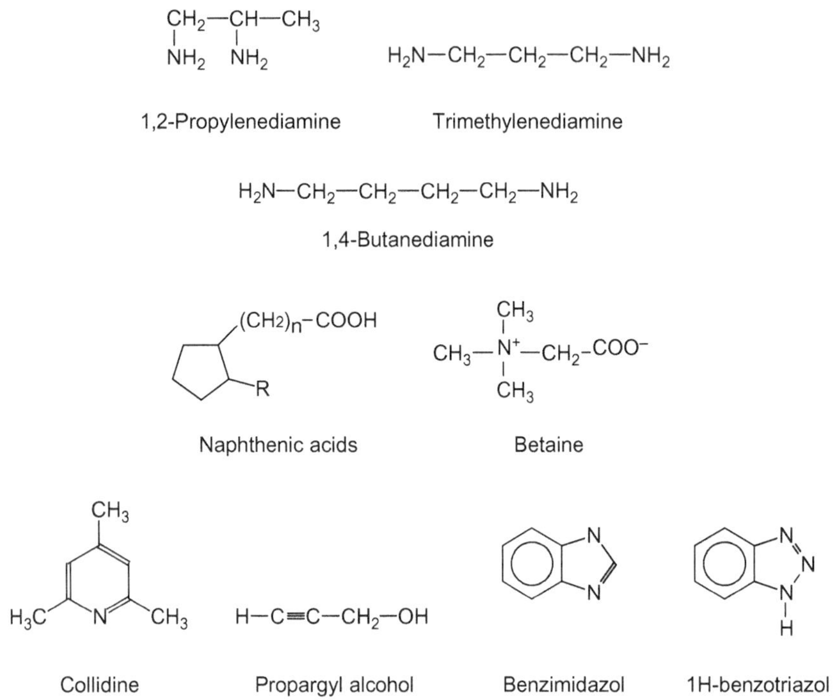

Figure 5.1 Diamines, acids alcohols, and azoles as corrosion inhibitors.

Table 5.4 Groups of corrosion inhibitors used in transportation

Group
Amides or imidazolines
Salts of nitrogenous molecules with carboxylic acids (fatty acids, naphthenic acids)
Nitrogen quaternaries
Polyoxylated amines, amides, and imidazolines
Nitrogen heterocyclics

Source: McKerrell EH, Lynes A. Development of an HPLC (high performance liquid chromatography) method for the determination of nitrogen containing corrosion inhibitors in a mixed hydrocarbon/glycol matrix. In: Proceedings Volume, no. 67. 3rd Royal Soc. Chem. Ind. Chem. in the Oil Ind. Int. Symp. (Manchester, England, 4/19–20/88); 1988. p. 212–22.

A product stream (40) contains the resultant compounds, including a large proportion of chlorine dioxide. The pressure of the entire system is maintained at a level sufficient to keep the chlorine dioxide dissolved in the reaction mixture in a product stream (40). A pressure gauge (42) may be located near a reactor (38) or at any other convenient location.

Table 5.5 Corrosion inhibitors

Inhibitor	**Reference**
Glutaraldehyde[a]	[47]
2-Phenylbenzimidazole and 2-methylbenzothiazole[b]	[48]
Benzotriazole and 1,3-dimethyl-2-thiourea	
Thiourea	
Sodium hexametaphosphate	
Tribasic sodium orthophosphate with polyurethane foam[c]	[49]

[a] Bacterial corrosion.
[b] Chemical corrosion.
[c] For external protection.

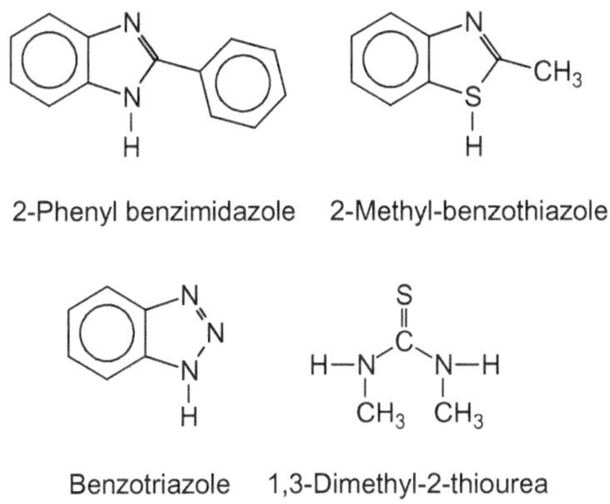

Figure 5.2 Corrosion inhibitors.

Normal pressures exceed 35 bar. The reactor (38) preferably operates at a pressure between 500 and 5000 psi so that it may be readily used for downstream injection, either directly or through a booster pump, into the water injection well with a typical temperature of about 55°C.

5.7.1 Organic Chemicals

A wide variety of mostly organic chemicals may be injected into pipeline fluids to reduce the corrosivity of the aqueous portion of the fluid stream [79, 80].

Typical families of types of film-forming inhibitors include substituted succinic anhydrides, imidazoline amines, and phosphate esters [81]. The adsorption mechanisms of some inhibitors have been studied (see Table 5.7).

Table 5.6 Corrosion inhibitors

Compound	References
Acetylenic alcohol[a]	[50]
Tall oil fatty acid anhydrides	[51, 52]
3-Phenyl-2-propyn-1-ol[b]	[53]
Dicyclopentadiene dicarboxylic acid salts[c]	[54, 55]
Hydroxamic acid	[56]
Cyclohexylammonium benzoate	[57, 58]
Acyl derivatives of tris(hydroxyethyl)perhydro-1,3,5-triazine	[59, 60]
2,4-Diamino-6-mercaptopyrimidine sulfate combined with oxysalts of vanadium, niobium, tantalum, titanium, zirconium, or hafnium	[61]
Aqueous alkanol amine solution[d]	[62, 63]
Quaternized fatty esters of alkoxylated alkylalkylene diamines	[64]
Mercaptoalcohols	[65]
Polysulfide[e]	[66]
Polyphosphonohydroxybenzene sulfonic acid compounds[f]	[67]
1-Hydroxyethylidene-1,1-diphosphonic acid[g]	[68]
2-Hydroxyphosphonoacetic acid[h]	[69]
Water-soluble 1,2-dithiol-3-thiones[i]	[70]
Sulfonated alkyl phenol[j]	[71]
Polythioether	[72]
Thiazolidines	[73]
Substituted thiacrown ethers pendent on vinyl polymers	[74]
Benzylsulfinylacetic acid or benzylsulfonylacetic acid	[75]
Halohydroxyalkylthio-substituted and dihydroxyalkylthio-substituted polycarboxylic acids[k]	[76]
Alkyl-substituted thiourea	[77]
2,5-Bis(*N*-pyridyl)-1,3,4-oxadiazoles	[78]

[a]In combination with ClO_2 treatment for bacteria control.
[b]Aqueous HCl.
[c]0.1–6% with antifreeze agents such as glycols.
[d]Gas stream containing H_2S or CO_2.
[e]Forms a film of iron disulfide
[f]Relatively nontoxic, substitution of chromate-based corrosion inhibitors, conventional phosphate, and organophosphonate inhibitors and the zinc-based inhibitors.
[g]CO_2 environment.
[h]Calcium chloride brine.
[i]10–500 ppm.
[j]5–200 ppm to inhibit naphthenic acid corrosion.
[k]In drilling equipment.

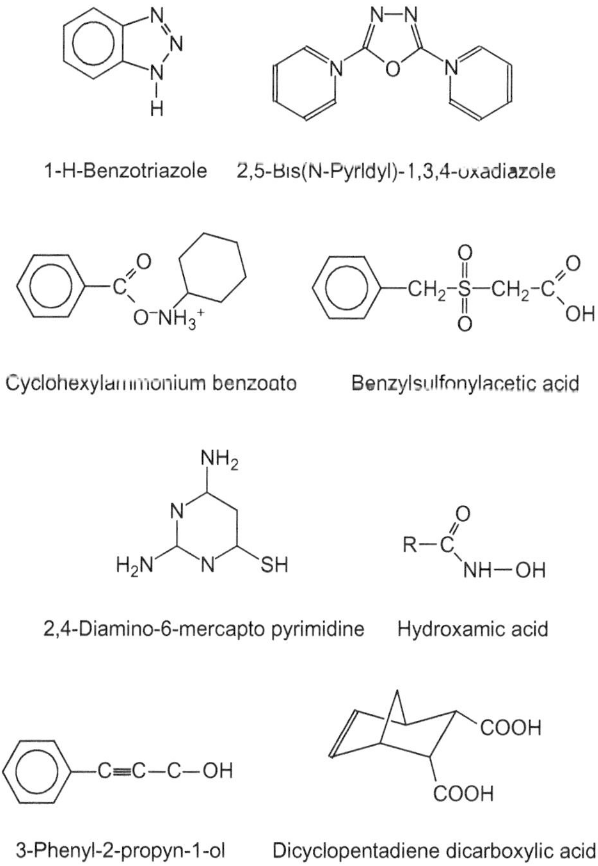

Figure 5.3 Miscellaneous corrosion inhibitors.

5.7.2 Oligomeric Amines

Derivates of alkynediols and polyalkylene polyamines can inhibit the corrosion of metal surfaces in pipelines because of the presence of a separate water phase in a pipeline transporting hydrocarbon fluids [82]. Exemplary polyamines are given in Table 5.8.

For example, an adduct was prepared from butynediol and pentaethylenehexamine with use of copper acetate as the catalyst [82]. Butynediols may be prepared by the condensation of acetylene with aldehydes or ketones in the presence of copper acetylide or related catalysts, such as malachite [83, 84].

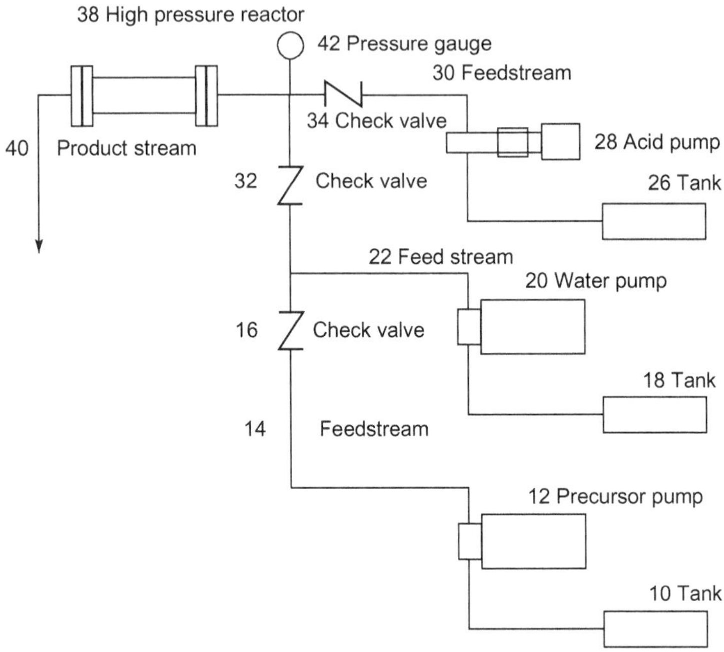

Figure 5.4 Process reactor system for acetylenic alcohol [50]. *(Source: Teeters SM. Corrosion inhibitor. US Patent 5084210, assigned to Chemlink Incorporated (Plano, TX); 1992. http://www.freepatentsonline.com/5084210.html).*

Table 5.7 Inhibitors tested for their adsorption mechanisms

Inhibitor	Chemisorption	Physisorption
Hexadecenylsuccinic anhydride	+	−
Hexadecylsuccinic anhydride	+	−
Octenylsuccinic anhydride	+	−
Benzyldimethylhexadecylammonium chloride	+	+
Trimethylhexadecylammonium chloride	+	+
Lauric acid	+	+
Sebacic acid	+	+
Cetylpyridinium chloride	+	+
Cetylamine	+	+
Stearylamine	+	+
Dimethyldodecylamine	+	+
Nalco/Exxon product	+	+
Cetylcyanide	+	+
Cetyl mercaptan	+	+
Dimethyldipalmitylammonium chloride	−	+
Folic acid	+	−

Source: Frenier WW, Wint DD. Multifaceted approaches for controlling top-of-theline corrosion in pipelines. Oil Gas Facil 2014;3(3):65–80. http://dx.doi.org/10.2118/169630-PA.

Table 5.8 Polyamines

Compound	
Ethylenediamine	Diethylenetriamine
Pentaethylenehexamine	Pentapropylenehexamine
Triheptylenediamine	

Source: Morin B, Goliaszewski AE. Methods for inhibiting water corrosion in crude oil pipelines. US Patent 5 643 500, assigned to BetzDearborn Inc. (Trevose, PA); 1997. http://www.freepatentsonline.com/5643500.html.

Another synthesis method according to [85] is shown in Eq. (iv).

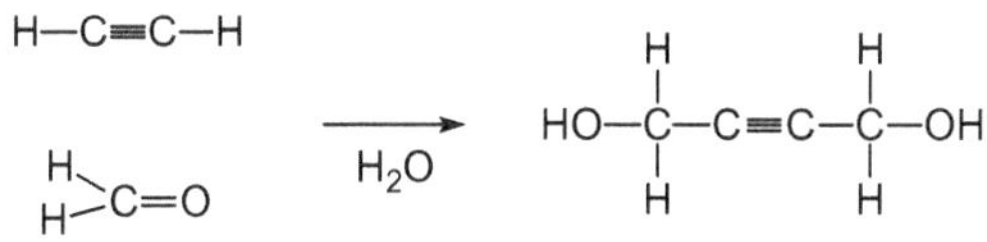

(iv)

This method has been claimed to provide an environmentally friendly, simple process, with high yield, and good economy. The synthesis of 3-hexyne-2,5-diol consists of the following steps [85]:

1. Preparation of an alumina-supported catalyst.
2. An aqueous solution of acetaldehyde and acetylene as raw materials in the alumina-supported catalyst. The reaction is conducted in an autoclave, and the molar ratio of acetylene to acetaldehyde is 2:1 to 3:1.
3. Alumina-supported catalyst and the weight ratio of acetaldehyde of 0.05:1 to 0.10:1 with a reaction temperature 30–100°C. N_2 is introduced and the reaction pressure is adjusted to 0.5–1.5 MPa. The reaction time is 3–10 h.
4. The reaction product obtained is filtered and the filtrate obtained after distillation is 3-hexyne-2,5-diol.

5.7.3 Foams

Foams may be used as inhibitor carriers [7]. A foam phase can be formed in pipelines, and various treatment chemicals, such as corrosion inhibitors, can be added to the foaming fluid and used to coat all pipeline surfaces. Some foam-inhibitor combinations are summarized in Table 5.9.

Table 5.9 Foam-inhibitor combinations

Foaming agent	Corrosion inhibitor
N-Decyl-*N*-dimethylamine oxide	3-Methoxyproplamine
Dodecylaminodipropionate	3-Methoxyproplamine bromide
Sodium C_{14}-C_{16} olefin sulfonate	Didecyldimethylammonium bromide
Dodecylbenzenesulfonic acid	Decylamine caprylate
Dentritic polymer	Dodecylbenzene octylamine caprylate

Source: Frenier WW, Wint DD. Multifaceted approaches for controlling top-of-the line corrosion in pipelines. Oil Gas Facil 2014;3(3):65–80. http://dx.doi.org/10.2118/169630-PA.

5.7.4 Oxygen Scavenger

Oxygen corrosion is often underestimated. Studies have shown that the corrosion can be limited when suitable oxygen scavengers are used. Hydrazine leads the group of chemicals that are available for oxygen removal. Because of its special properties, it is used for corrosion control in heating systems and in drilling operations, well workover, and cementing [86].

5.7.5 Hydrogen Sulfide Removal

Hydrogen sulfide is produced by these sulfate reducing bacteria and is released to the protected environment, where it reacts with the dissolved iron from the corrosion process to form iron sulfide [87]. So it is sometimes necessary to remove hydrogen sulfide from a drilling mud. Techniques using iron compounds that form sparingly soluble sulfides have been developed—for example, with iron(II) oxalate and iron sulfate [88]. The sulfur is precipitated as FeS. Ferrous gluconate is an organic iron-chelating agent that is stable at pH levels as high as 11.5 [89].

Zinc compounds have high reactivity with regard to H_2S and therefore are suitable for the quantitative removal of even small amounts of hydrogen sulfide [90]. However, at high temperatures they may negatively affect the rheology of drilling fluids.

5.8 INHIBITORS FOR SPECIAL TASKS

5.8.1 Inhibitors for Aqueous Media

Corrosion of metal surfaces in an aqueous medium has long been a problem for the oil and gas industries [91]. During the production of oil and gas several other corrosive components are present, such as brines, organic acids, carbon dioxide, hydrogen sulfide, and microorganisms.

These aggressive constituents can cause severe corrosion of metal pipes, which are often made of low-alloy steels. This problem is even more troublesome in deep-sea operations, where replacement of corroded equipment is difficult and costly.

Among the most frequently used corrosion inhibitors in crude oil and natural gas extraction are imidazoline derivatives. Corrosion inhibitors for use in an aqueous medium are shown in Fig. 5.5.

5.8.2 Iron Sulfide

Ferrous sulfide deposits are a major source of economic loss in the oil industry. The deposits are mainly the result of a reaction between hydrogen sulfide, formed by sulfate-reducing bacteria, and ferrous metal oilfield equipment or an iron compound in the formation.

They obstruct the flow of oil through wells and in the adjacent strata and also in pipelines and in the processing and refinery plant. Ferrous sulfide particles also tend to stabilize oil-water emulsions which often form, especially during secondary oil recovery, and present major problems to oil producers.

The simplest way to dissolve a deposit of ferrous sulfide is by contact with strong acid. Unfortunately this method generates large volumes of highly toxic hydrogen sulfide gas, which in the past has been responsible for fatalities.

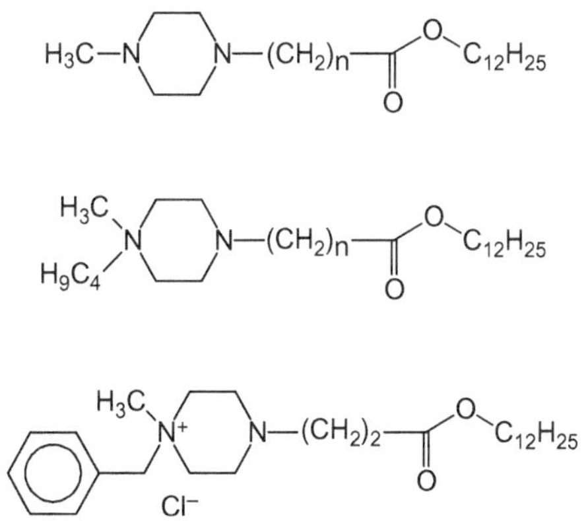

Figure 5.5 Corrosion inhibitors for aqueous media. *(Source: Acosta EJ, Clark JC. Corrosion inhibitors for an aqueous medium. US Patent 8 105 987, assigned to Nalco Company (Naperville, IL); 2012. http://www.freepatentsonline.com/8105987.html).*

An alternative method for treating the deposits with powerful oxidizing agents avoids the toxicity hazards but produces oxidation products, including elemental sulfur, which are so corrosive to pipework that it has not generally been practiced.

It has been found that tris(hydroxyorgano)phosphines (THPs) are capable of solubilizing iron sulfide by forming a bright red water-soluble complex. THP is believed to be formed in oil wells treated with tetrakis(hydroxmethyl)phosphonium salts. Such salts, especially the sulfate salts, are commonly added to oil wells as biocides.

Such salts are highly effective at killing the sulfate-reducing bacteria, whose activity was largely responsible for the original formation of the iron sulfide deposits. However, the effectiveness of THP as a solubilizing agent for iron sulfides varies considerably from well to well. This arises because the complex with iron sulfide requires the presence of ammonium ions. Although ammonium ions are normally present in oilfield water, the concentration is frequently less than the optimum for iron sulfide removal. The pH is critical to the formation of the complex.

Water-soluble condensates of THP with co-condensable organic nitrogen compounds such as urea and thiourea are also capable of solubilizing iron sulfide and provide more consistent performance than THP-ammonia mixtures. However, such condensates may also cause deposition of polymers if they are used in high concentration.

THP and amino carboxylic acids or amino phosphonic acids act synergistically to dissolve iron sulfide deposits, even in the absence of ammonia. Moreover, THP is stable in the presence of amino phophonates even when the two are formulated together and stored for extended periods before use [92].

5.9 EROSION

Erosion phenomena are common in the petroleum industry. Damage of material occurs frequently in high-pressure pipelines in the course of hydraulic fracturing [93].

With increasing operation times, the erosion and corrosion defects on the inner surface of the pipeline result in serious material loss and thus in a failure of the equipment.

A device to simulate the erosive wear behavior of metal materials caused by a fracturing fluid has been developed [93]. The erosion failure mechanism caused by various parameters, such as the velocity of multiphase flow, fracturing proppant, and impact angles, has been investigated. Also, microcosmic surface testing was used to analyze the erosion failure mechanism of metal materials in high-pressure pipelines.

REFERENCES

[1] Baboian R, Hack HP, Efird KD. Petroleum. In: Baboian R, editor. Corrosion tests and standards: application and interpretation. 2nd ed. West Conshohocken, PA: ASTM International; 2005. p. 425–34 [chapter 64]. ISBN 9780803120983, http://dx.doi.org/10.1520/MNL11042M.

[2] Baboian R, Dean SW, Rothman PS, Young WT. Pipeline. In: Baboian R, editor. Corrosion tests and standards: application and interpretation. 2nd ed. West Conshohocken, PA: ASTM International; 2005. p. 696–706 [chapter 64]. ISBN 9780803120983, http://dx.doi.org/10.1520/MNL11070M.

[3] Popoola L, Grema A, Latinwo G, Gutti B, Balogun A. Corrosion problems during oil and gas production and its mitigation. Int J Ind Chem 2013;4(1):35. http://dx.doi.org/10.1186/2228-5547-4-35.

[4] Groysman A. Corrosion in systems for storage and transportation of petroleum products and biofuels: identification, monitoring and solutions. Berlin: Springer; 2014. ISBN 9789400778832.

[5] Fang BY, Atrens A, Wang JQ, Han EH, Zhu ZY, Ke W. Review of stress corrosion cracking of pipeline steels in "low" and "high" pH solutions. J Mater Sci 2003;38(1):127–32. http://dx.doi.org/10.1023/A:1021126202539.

[6] Nanninga N, Slifka A, Levy Y, White C. A review of fatigue crack growth for pipeline steels exposed to hydrogen. J Res Natl Inst Standards Tech 2010;115(6):437–52. http://nvlpubs.nist.gov/nistpubs/jres/115/6/V115.N06.A04.pdf.

[7] Frenier WW, Wint DD. Multifaceted approaches for controlling top-of-the-line corrosion in pipelines. Oil Gas Facil 2014;3(3):65–80. http://dx.doi.org/10.2118/169630-PA.

[8] Nalli K. Corrosion and its mitigation in the oil and gas industries. In: Holloway MD, Nwaoha C, Onyewuenyi OA, editors. Process plant equipment: operation, control, and reliability. Hoboken, NJ: John Wiley & Sons, Inc.; 2012. p. 673–79 [Appendix VI]. ISBN 9781118162569, http://dx.doi.org/10.1002/9781118162569.app6.

[9] Dunlop AK, Hassell HL, Rhodes PR. Fundamental considerations in sweet gas well corrosion. In: Corrosion 83. Houston: National Association of Corrosion Engineers; 1983.

[10] Frankel GS. Pitting corrosion of metals. J Electrochem Soc 1998;145(6):2186. http://dx.doi.org/10.1149/1.1838615.

[11] Fisher LE. Corrosion inhibitors and neutralizers: past, present and future. In: Proceedings Volume. Annu. NACE Corrosion Conf. (Corrosion 93) (New Orleans, 3/7–12/93); 1993.

[12] Becker JR. Corrosion and scale handbook. Tulsa, OK: Pennwell Publishing Co; 1998. ISBN 0-87814-749-7.

[13] ASTM G 205-10. Standard guide for determining corrosivity of crude oils. ASTM Standard, Book of Standards, Vol. 3.02 ASTM G 205-10. West Conshohocken, PA: ASTM International; 2010. http://www.astm.org/Standards/G205.htm.

[14] ASTM G 170-06. Standard guide for evaluating and qualifying oilfield and refinery corrosion inhibitors in the laboratory. ASTM Standard, Book of Standards, Vol. 3.02 ASTM G 170-06, West Conshohocken, PA: ASTM International; 2012. http://www.astm.org/Standards/G170.htm.

[15] NACE. Wheel test method used for evaluation of film-persistent corrosion inhibitors for oilfield applications. Test 1D182. Houston, TX: NACE International; 2005. http://www.nace.org/cstm/Store/Product.aspx?id=205ebbf5-c59c-4cf4-8451-e1df6fff02df.
[16] Sivokon IS, Andreev NN. Laboratory assessment of the efficiency of corrosion inhibitors at oilfield pipelines of the West Siberia region I: objective setting. Int J Corros Scale Inhib 2012;1(1):65–79. http://ijcsi.pro/files/2012/09/ijcsi-2012-1-1-6-65-79.pdf.
[17] Sivokon IS, Vershok DB, Andreev NN. Laboratory assessment of corrosion inhibitors effectiveness at oilfield pipelines of West Siberian region. II. Tests in a U-cell. Int J Corros Scale Inhib 2012;1(2):130–45. http://ijcsi.pro/files/2012/10/ijcsi-2012-1-2-6-130-145.pdf.
[18] Andreev NN, Anufriev NG, Sivokon IS. Laboratory assessment of corrosion inhibitors effectiveness at oilfield pipelines of West Siberian region. III. Bubble test. Int J Corros Scale Inhib 2013;2(1):17–29. http://ijcsi.pro/files/2013/Issue_1/ijcsi-2013-v2-n1-p3-pp17-29.pdf.
[19] Sivokon IS, Makarychev YB, Kuzenkov YA, Andreev NN. Laboratory assessment of corrosion inhibitors effectiveness at oilfield pipelines of West Siberian region. IV. Tests in a flow recirculation loop. Int J Corros Scale Inhib 2013;2(3):203–15. http://ijcsi.pro/files/2013/Issue_3/ijcsi-2013-v2-n3-p3-pp203-215.pdf.
[20] Gedvillo IA, Oleinik SV, Sivokon IS, Andreev NN. Laboratory assessment of the efficiency of corrosion inhibitors at oilfield pipelines of the West Siberia region V. Rotating cylinder and cage. Int J Corros Scale Inhib 2013;2(4):287–303. http://ijcsi.pro/files/2013/Issue_4/ijcsi-2013-v2-n4-p4-pp287-303.pdf.
[21] Obuka NSP, Okoli CN, Ikwu GOR, Chukwumuanya EO. Review of corrosion kinetics and thermodynamics of CO_2 and H_2S corrosion effects and associated prediction/evaluation on oil and gas pipeline system. Int J Scient Techn Res 2012;1:156–62. http://www.ijstr.org/final-print/may2012/Review-of-corrosion-kinetics-and-thermodynamics-of-co2-and-h2s-corrosion-effects-and-associated-prediction-evaluation-on-oil-and-gas-pipeline-system.pdf.
[22] Roberge P. Corrosion basics: an introduction. Houston, TX: NACE International; 2006. ISBN 9781575901985.
[23] Boreman DJ, Wimmer BO, Leewis KG. Repair technologies for gas transmission pipelines. Pipeline Gas J 2000;227(3):46–51.
[24] Muthukumar N, Rajasekar A, Ponmariappan S, Mohanan S, Maruthamuthu S, Muralidharan S, et al. Microbiologically influenced corrosion in petroleum product pipelines: a review. Ind J Exp Biol 2003;41(9):1012–22. http://nopr.niscair.res.in/handle/123456789/17162.
[25] Pope DH. Concern over MIC (microbiologically-influenced corrosion) expanding among corrosion engineers. Pipe Line Gas Ind 1997;80(2):23–5.
[26] Rajasekar A, Anandkumar B, Maruthamuthu S, Ting YP, Rahman P. Characterization of corrosive bacterial consortia isolated from petroleum-product-transporting pipelines. Appl Microbiol Biotechnol 2010;85(4):1175–88. http://dx.doi.org/10.1007/s00253-009-2289-9.
[27] Rajasekar A, Ponmariappan S, Maruthamuthu S, Palaniswamy N. Bacterial degradation and corrosion of naphtha in transporting pipeline. Curr Microbiol 2007;55(5):374–81. http://dx.doi.org/10.1007/s00284-007-9001-z.
[28] Wu Y. Chemical problems and costs involved in downstream gas systems. In: Proceedings Volume. NACE Int. Corrosion Forum (Corrosion 90) (Las Vegas, 4/23–27/90); 1990.
[29] Sartori G, Savage DW, Dalrymple DC, Ballinger BH, Blum SC, Wales WE. Esterification of acidic crudes. US Patent 6 251 305, assigned to Exxon Research and Engineering Company (Annandale, NJ); 2001. http://www.freepatentsonline.com/6251305.html.

[30] Johnson JD, Fu SL, Bluth MJ, Marble RA. Enhanced corrosion protection by use of friction reducers in conjunction with corrosion inhibitors. US Patent 5 939 362, assigned to Nalco/Exxon Energy Chemicals, LP (Sugar Land, TX); 1999. http://www.freepatentsonline.com/5939362.html.
[31] McMahon AJ, Smith PS, Lee Y. Drag reducing chemical enables increased sea water injection without increasing the oxygen corrosion rate. In: Proceedings Volume. NACE Int. Corrosion Conf. (Corrosion 97) (New Orleans, 3/9–14/97); 1997.
[32] Krivohlavek D, Ziesmer RR. Universal cross linking compound and polymer. US Patent 6 451 886, assigned to Krivohlavek and Dennis (Claremore, OK); 2002. http://www.freepatentsonline.com/6451886.html.
[33] Buchheit RG, Hinkebein TE, Hlava PF, Melton DG. The effects of latex additions on centrifugally cast concrete for internal pipeline protection. In: Proceedings Volume, vol. 4. 12th SPE/nace Int. Corrosion Congr. (Houston, 9/19–24/93); 1993. p. 2854–64. ISBN 1-877914-65-7.
[34] Osborne D. The development in the use of fre (fiber reinforced epoxy) pipe systems for industrial and offshore applications. In: Proceedings Volume, no. 7. 10th Bhr Group Ltd et al Pipe Protect Int. Conf. (Amsterdam, Netherlands, 11/10–12/93); 1993. p. 27–45. ISBN 0-82598-875-3.
[35] Miyajima Y, Kariyazono Y, Funatsu S, Endo E. Durability of polyethylene coated steel pipe at elevated temperature. In: Proceedings Volume, no. 7. 10th Bhr Group Ltd et al Pipe Protect Int. Conf. (Amsterdam, Netherlands, 11/10–12/93); 1993. p. 183–90. ISBN 0-82598-875-3.
[36] Arai T, Ohkita M, Application of polypropylene coating system to pipeline for high temperature service. In: Proceedings Volume. 8th Bhra Internal & External Protect of Pipes Int. Conf. (Florence, Italy, 10/24–26/89); 1989. p. 189–201. ISBN 0-87201-022-8.
[37] Palmer FH, Wright A. Polyurethanes as insulating systems for high temperature marine pipelines. In: Proceedings Volume. Bhr Group Ltd et al Pipe Protect Conf. (Cannes, France, 9/23–25/91); 1991. p. 169–87. ISBN 1-85166-713-X.
[38] Kenworthy T. Polyurea coatings for offshore oil rigs in the Gulf of Mexico. Mater Perf 2001;40(10):40–2.
[39] McGarvey GB, Falkiner RJ, Slim DR, Knickerbocker BM. Corrosion control in acid gas removal equipment by the situ generation of polysulfide ions. US Patent 8 916 117, assigned to ExxonMobil Research and Engineering Company (Annandale, NJ); 2014. http://www.freepatentsonline.com/8916117.html.
[40] Ryan JM, McCann P, O'Brrian J, Propane recovery, Ryan Holmes process. In: McKetta J, editor. Encyclopedia of chemical processing and design. New York: Marcel Dekker; 1976. ISBN 9780824724955.
[41] Kohl A, Nielson R. Gas purification. 5th ed. Houston, TX: Gulf Pub.; 1997. ISBN 9780080507200.
[42] Madden II PC, Foroulis ZA. Use of thiosulfate salt for corrosion inhibition in acid gas scrubbing processes. US Patent 4 944 917, assigned to Exxon Research and Engineering Company (Florham Park, NJ); 1990. http://www.freepatentsonline.com/4944917.html.
[43] Madden II PC. Use of sulfur dioxide for corrosion inhibition in acid gas scrubbing processes. US Patent 4 857 283, assigned to Exxon Research and Engineering Company (Florham Park, NJ); 1989. http://www.freepatentsonline.com/4857283.html.
[44] Siskin M, Mozeleski EJ, Fedich RB, Wang FCY. Alkylamino alkyloxy (alcohol) monoalkyl ether for acid gas scrubbing process. US Patent 8 486 183, assigned to ExxonMobil Research and Engineering Company (Annandale, NJ); 2013. http://www.freepatentsonline.com/8486183.html.
[45] Behm M, Simonsson D. Electrochemical production of polysulfides and sodium hydroxide from white liquor: Part I: Experiments with rotating disc and ring-disc

electrodes. J Appl Electrochem 1997;27(5):507–18. http://dx.doi.org/10.1023/A:1018486309601.
[46] Dietsche F, Essig M, Friedrich R, Kutschera M, Schrepp W, Witteler H, et al. Organic corrosion inhibitors for interim corrosion protection. In: CORROSION 2007. Nashville, TN: NACE International; 2007. p. 1–12. https://www.onepetro.org/download/conference-paper/NACE-07358?id=conference-paper%2FNACE-07358.
[47] El-Gamal IM, Gad EAM. Low temperature rheological behavior of umbarka waxy crude and influence of flow improver. Rev Inst Franc Petrol 1997;52(3):369–79.
[48] Koshel KC, Bhatia JS, Kumar S, Samant AK, Corrosion problem in Kalol injection water pipeline system and its control by using corrosion inhibitors. ONGC Bull 1988;25(2):115–33.
[49] Bohon WM, Ruschau GR. Method for inhibiting external corrosion on an insulated pipeline. US Patent 6 273 144, assigned to Atlantic Richfield Company (Los Angeles, CA); 2001. http://www.freepatentsonline.com/6273144.html.
[50] Teeters SM. Corrosion inhibitor. US Patent 5 084 210, assigned to Chemlink Incorporated (Plano, TX); 1992. http://www.freepatentsonline.com/5084210.html.
[51] Fischer ER, Parker III JE. Tall oil fatty acid anhydrides as corrosion inhibitor intermediates. In: Proceedings Volume. 50th Annu. NACE Int. Corrosion Conf. (Corrosion 95) (Orlando, FL, 3/26–31/95); 1995.
[52] Fischer ER, Parker III JE. Tall oil fatty acid anhydrides as corrosion inhibitor intermediates. Corrosion 1997;53(1):62–4.
[53] Growcock FB, Lopp VR. The inhibition of steel corrosion in hydrochloric acid with 3-phenyl-2-propyn-1-ol. Corrosion Sci 1988;28(4):397–410.
[54] Darden JW, McEntire EE. Dicyclopentadiene dicarboxylic acid salts as corrosion inhibitors. EP Patent 0 200 850 assigned to Texaco Development Corporation; 1986. https://www.google.at/patents/EP0200850A1?cl=en.
[55] Darden JW, McEntire EE. Dicyclopentadiene dicarboxylic acid salts as corrosion inhibitors. US Patent 4 592 853, assigned to Texaco Inc. (White Plains, NY); 1986. http://www.freepatentsonline.com/4592853.html.
[56] Fong DW, Khambatta BS. Hydroxamic acid containing polymers used as corrosion inhibitors. US Patent 5 308 498, assigned to Nalco Chemical Company (Naperville, IL); 1994. http://www.freepatentsonline.com/5308498.html.
[57] Johnson DM, Ippolito JS. Corrosion inhibitor and sealable thread protector end cap for tubular goods. US Patent 5 352 383, assigned to Centrax International Corp. (Houston, TX); 1994. http://www.freepatentsonline.com/5352383.html.
[58] Johnson DM, Ippolito JS. Corrosion inhibitor and sealable thread protector end cap for tubular goods. US Patent 5 452 749, assigned to Centrax International Corp. (Houston, TX); 1995. http://www.freepatentsonline.com/5452749.html.
[59] Au AT. Acyl derivatives of tris-hydroxy-ethyl-perhydro-1,3,5-triazine. US Patent 4 605 737, assigned to The Dow Chemical Company (Midland, MI); 1986. http://www.freepatentsonline.com/4605737.html.
[60] Au AT, Hussey HF. Method of inhibiting corrosion using perhydro-s-triazine derivatives. US Patent 4 830 827, assigned to The Dow Chemical Company (Midland, MI); 1989. http://www.freepatentsonline.com/4830827.html.
[61] Ramanarayanan TA, Vedage HL. Inorganic/organic inhibitor for corrosion of iron containing materials in sulfur environment. US Patent 5 279 651, assigned to Exxon Research & Engineering Company (Florham Park, NJ); 1994. http://www.freepatentsonline.com/5279651.html.
[62] Schutt HU. Reducing stress corrosion cracking in treating gases with alkanol amines. US Patent 4 959 177, assigned to Shell Oil Company (Houston, TX); 1990. http://www.freepatentsonline.com/4959177.html.

[63] Veldman RR, Trahan DO. Gas treating solution corrosion inhibitor. US Patent 6 059 992; 2000. http://www.freepatentsonline.com/6059992.html.
[64] Wirtz H, Hoffmann H, Ritschel W, Hofinger M, Mitzlaff M, Wolter D. Esters of fatty acids, which may be quaternized, from oxalkylated alkylalkylenediamines. US Patent 4 997 912, assigned to Hoechst Aktiengesellschaft (Frankfurt am Main, DE); 1991. http://www.freepatentsonline.com/4997912.html.
[65] Ahn YS, Jovancicevic V. Mercaptoalcohol corrosion inhibitors. US Patent 6 645 399, assigned to Baker Hughes Incorporated (Houston, TX); 2003. http://www.freepatentsonline.com/6645399.html.
[66] Gay RJ, Gay CC, Matthews VM, Gay FEM, Chase V. Dynamic polysulfide corrosion inhibitor method and system for oil field piping. US Patent 5 188 179; 1993. http://www.freepatentsonline.com/5188179.html.
[67] Kreh RP. Method of inhibiting corrosion and scale formation in aqueous systems. US Patent 5 073 339, assigned to W.R. Grace & Co.—Conn (Columbia, MD); 1991. http://www.freepatentsonline.com/5073339.html.
[68] Sekine I, Yuasa M, Shimode T, Takaoka K. Inhibition of corrosion; 1991.
[69] Zefferi SM, May RC. Corrosion inhibition of calcium chloride brine. US Patent 5 292 455, assigned to Betz Laboratories, Inc. (Trevose, PA); 1994. http://www.freepatentsonline.com/5292455.html.
[70] Oude Alink BA. Water soluble 1,2-dithio-3-thiones. US Patent 5 252 289, assigned to Petrolite Corporation (St. Louis, MO); 1993. http://www.freepatentsonline.com/5252289.html.
[71] Babaian-Kibala E. Naphthenic acid corrosion inhibitor. US Patent 5 252 254, assigned to Nalco Chemical Company (Naperville, IL); 1993. http://www.freepatentsonline.com/5252254.html.
[72] Incorvia MJ. Polythioether corrosion inhibition system. US Patent 4 759 908, assigned to Texaco Inc. (White Plains, NY); 1988. http://www.freepatentsonline.com/4759908.html.
[73] Alink BAMO, Outlaw BT. Thiazolidines and use thereof for corrosion inhibition. US Patent 6 592 781, assigned to Baker Hughes Incorporated (Houston, TX); 2003. http://www.freepatentsonline.com/6592781.html.
[74] Minevski LV, Gaboury JA. Thiacrown ether compound corrosion inhibitors for alkanolamine units. US Patent 6 187 227, assigned to Betzdearborn (Trevose, PA); 2001. http://www.freepatentsonline.com/6187227.html.
[75] Lindstrom MR, Mark HW. Inhibiting corrosion: benzylsulfinylacetic acid or benzylsulfonylacetic acid. US Patent 4 637 833, assigned to Phillips Petroleum Co. (Bartlesville, OK); 1987. http://www.freepatentsonline.com/4637833.html.
[76] Lindstrom MR, Louthan RP. Inhibiting corrosion. US Patent 4 670 163, assigned to Phillips Petroleum Company (Bartlesville, OK); 1987. http://www.freepatentsonline.com/4670163.html.
[77] Tang Y, Han Z, Wang H, Chen H. Sp-2 acid corrosion inhibitor. J Univ Pet China 1995;19(1):98–101.
[78] Bentiss F, Lagrenee M, Traisnel M. 2,5-bis(N-pyridyl)-1,3,4-oxadiazoles as corrosion inhibitors for mild steel in acidic media. Corrosion 2000;56(7):733–42.
[79] Palmer JW, Hedges W, Dawson JL. The use of corrosion inhibitors in oil and gas production. European Federation of Corrosion Publications. Leeds: Maney Pub.; 2004. http://public.eblib.com/EBLPublic/PublicView.do?ptiID=677842.
[80] Kelland M. Production chemicals for the oil and gas industry. 2nd ed. Hoboken, NJ: Taylor and Francis; 2014. ISBN 9781439873816.
[81] Frenier W, Ziauddin M. Chemistry for enhancing the production of oil and gas. Richardson, TX: Society of Petroleum Engineers; 2014. ISBN 9781613993170.

[82] Morin B, Goliaszewski AE. Methods for inhibiting water corrosion in crude oil pipelines. US Patent 5 643 500, assigned to BetzDearborn Inc. (Trevose, PA); 1997. http://www.freepatentsonline.com/5643500.html.
[83] Coyne DM, Riggs JOL. Corrosion inhibition. US Patent 3 211 667, assigned to Continental Oil Co.; 1965. https://www.google.com/patents/US3211667.
[84] Fremont JM. Preparation of butynediol. US Patent 4 584 418, assigned to E.I. Du Pont de Nemours and Company (Wilmington, DE); 1986. http://www.freepatentsonline.com/4584418.html.
[85] Hui L, Tian WJ, Zhang W. Synthesis method of 3-hexyne-2,5-diol. CN Patent 8 603 325, assigned to Guangdong Electronic Senior Technical School; 2013. https://www.google.at/patents/CN102285867B?cl=en, cN Patent 102,285,867.
[86] Sikora D. Hydrazine—a universal oxygen scavenger (hydrazyna—uniwersalny inhibitor korozji tlenowej w pluczkach wiertniczych). Nafta Gaz (Pol) 1994;50(4):161–8.
[87] Martin RL, Brock GF, Dobbs JB. Corrosion inhibitors and methods of use. US Patent 6 866 797, assigned to BJ Services Company (Houston, TX); 2005. http://www.freepatentsonline.com/6866797.html.
[88] Sunde E, Olsen H. Removal of H_2S in drilling mud. US Patent 6 365 053, assigned to Statoil Asa (Stavanger, NO); 2002. http://www.freepatentsonline.com/6365053.html.
[89] Davidson E. Method and composition for scavenging sulphide in drilling fluids and composition. US Patent 6 746 611, assigned to Halliburton Energy Services, Inc. (Duncan, OK); 2004. http://www.freepatentsonline.com/6746611.html.
[90] Wegner C, Reichert G. Hydrogen sulfide scavenger in drilling fluids (schwefelwasserstoff-scavenger in bohrspülungen). In: Proceedings Volume. BASF et al Chem. Prod. in Petrol. Prod. Mtg. H_2S—A Hazardous Gas in Crude Oil Recovery Discuss (Clausthal-Zellerfeld, Ger, 9/12–13/90); 1990.
[91] Acosta EJ, Clark JC. Corrosion inhibitors for an aqueous medium. US Patent 8 105 987, assigned to Nalco Company (Naperville, IL); 2012. http://www.freepatentsonline.com/8105987.html.
[92] Fidoe SD, Talbot RE, Jones CR, Gabriel R. Treatment of iron sulphide deposits. US Patent 6 926 836, assigned to Rhodia Consumer Specialties Limited (Watford, GB); 2005. http://www.freepatentsonline.com/6926836.html.
[93] Zhang J, Fan J, Xie Y, Wu H. Research on erosion of metal materials for high pressure pipelines. Adv Mater Res (Zur) 2012;482-484(2):1592–5. http://dx.doi.org/10.4028/www.scientific.net/AMR.482-484.1592.

CHAPTER 6

Drag Reduction and Flow Improvement

Drag-reducing additives are important in oil drilling applications and the maintenance of pumping equipment in pipelines. Flow drag in pipes can be reduced by addition of a small amount of high molecular weight polymer. Pipeline flow improvers, or drag-reducing additives, have been used in the petroleum industry for many years [1].

6.1 HISTORY

Pipeline flow improvers, or drag-reducing agents, have been used in the petroleum industry for many years [1]. The first application of drag reducers in the petroleum industry was to reduce the downhole pressure loss during the pumping of the fluids downhole to fracture-tight formations.

The first application of drag reducers was the use of guar in oil well fracturing, presently a routine practice. The quantities of the drag-reducing additives used in this application were very large. One of the first large-scale pipeline applications was to increase the throughput of crude oil in the Trans-Alaskan pipeline in 1979 with oil-soluble polymers. These agents were highly successful in applying a drag-reducing phenomenon.

Since then, the use of drag-reducing additives has increased in refined product pipelines. This is justified by an economic factor—namely, to offset power costs. The cost of moving an additional barrel of product through a pipeline system can be less than $0.05 per barrel. This low cost has been made possible by improved performance of commercially available drag reducers and a nearly unchanging price structure.

In the subsea production of oil and gas, production piping typically presents a significant bottleneck because of the difficulty and expense associated with the subsea installation of the piping. The production decrease caused by bottlenecking of subsea flow lines can have severe economic ramifications owing to the resulting inability to run the hydrocarbon production system at full capacity [2].

Guide to the Practical Use of Chemicals in Refineries and Pipelines
http://dx.doi.org/10.1016/B978-0-12-805412-3.00006-4

6.2 OPERATING COSTS

Pipeline operators of liquid hydrocarbon products can decrease operating costs by using a drag-reducing flow improver to eliminate the need for underutilized intermediate or booster pump stations [3, 4]. Product lines operating below their capacity or that use boosters only intermittently can also realize cost savings.

The overall benefits are likely to be most significant in 6–8-in lines operating between 67% and 92% of their rated throughput capacity. Using computer modeling techniques, engineers have demonstrated potential power savings of up to 22% (from lower demand charges and reduced energy use) for systems using booster stations 85% of the operating time. When stations operate only 70% of the time, total energy cost savings can approach 35%, depending on the diameter of the line and electricity costs.

6.3 GENERAL THEORETICAL ASPECTS

The effects of various parameters on pressure drop reduction caused by addition of small amounts of drag-reducing polymers have been investigated in crude oil pipelines [5]. The results obtained indicate that the amount of drag reduction increases with temperature, oil flow rate, pipe roughness, and the concentration of the additive.

By investigation of the various experimental parameters, a general model has been proposed for predicting the phenomenon of drag reduction in crude oil pipelines [6]. The proposed model uses a comprehensive analysis of various operation parameters, such as oil flow rate and temperature, pipe diameter and roughness, and various types of drag-reducing agents and their concentration.

The friction factor has been calculated from experimental data and the relevant model was developed by consideration of the changes of the friction factor characterized by the Reynolds number. The proposed model shows good agreement with the experimental data.

6.4 CLASSES OF DRAG REDUCERS

Because of the reduction of the apparent viscosity, drag reducers are useful for saving energy required for pumping. Some drag reducers are given in Table 6.1 and Fig. 6.1.

Table 6.1 Drag reducers

Chemicals	**References**
Low-density polyethylene[a]	[7]
Copolymer of a linear α-olefin with cross-linkers[b]	[8]
Polyacrylamides[c]	[9–11]
Poly(alkylene oxide)[c]	[9–11]
Fluorocarbons[d]	[12]
Poly(alkyl methacrylate)s[e]	[13, 14]
Terpolymer of styrene, alkyl acrylate, and acrylic acid or methacrylic acid[f]	[15]

[a] Olefin up to 10 mol%.
[b] α-Olefins are 1-hexene, 1-octene, 1-decene, and 1-dodecene; cross-linkers are divinylbenzene or organosiloxanes with pendent vinyl groups.
[c] Water-soluble drag reducers for emulsions.
[d] For asphaltenic crude oils.
[e] Esters with C_{10} to C_{18} and ionic monomers; reduce friction in the flow of hydrocarbons by a factor of 5 at concentrations of 25 ppm.
[f] Styrene also includes *tert*-butylstyrene (drag reducer for hydrocarbon fluids).

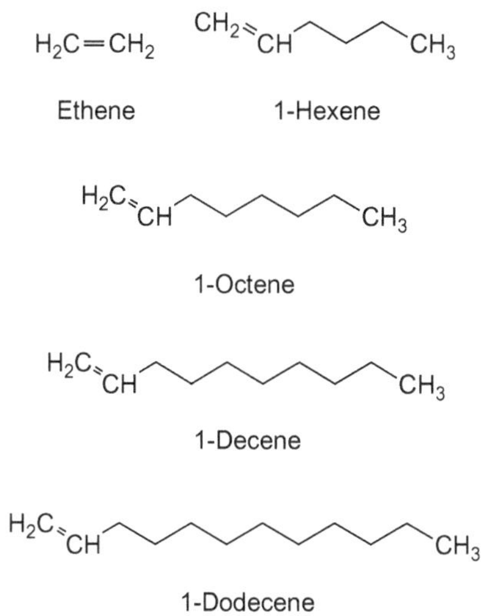

Figure 6.1 Monomers for polyolefins as drag reducers.

6.5 MECHANISM OF DRAG REDUCERS

A review on drag-reducing polymers is given in the literature [16, 17]. It has been suggested that drag reduction occurs by the interactions between elastic macromolecules and macrostructures in turbulent flow. In turbulent pipe flow, the region near the wall, composed of a viscous sublayer and a buffer layer, plays a major role in drag reduction.

The most serious problem in the effectiveness of drag reducers is the chain degradation of polymers by shear strains in turbulent flow. Ultra high molecular weight polymers are more sensitive to shear-induced degradation [18]. Polymers with linear-chain structures are more vulnerable than branched polymers [19] and natural gums with semirigid structures [20].

The mechanism of shear degradation is assumed to be associated with chain elongation. The chain degradation is often observed when the shear rate is increased to a critical point, after which drag reduction sharply decreases.

The friction drag and heat transfer reduction phenomena associated with turbulent flows of so-called drag-reducing fluids are not well understood [21]. It is believed that elastic fluid properties are strongly related to these phenomena. However, not all drag-reducing fluids are viscoelastic, nor are all viscoelastic fluids drag reducing, suggesting that drag reduction and viscoelasticity are probably incidentally accompanying phenomena.

It is argued that turbulence suppression (ie, flow laminarization) is a determining factor for the reduction phenomena, but not the fluid elasticity, because of the flow-induced anisotropic fluid structure and the associated properties, whereas the fluid elasticity may be a major cause of the laminar heat transfer augmentation.

6.5.1 Alternatives to Polymer Additives

Recently, the use of polymer additives for drag reduction has been criticized [22]: the conventional mainstream method of the use of polymer additives to reduce drag in pipes is disadvantageous from certain environmental perspectives.

The widespread preference for the use of polymeric and surfactant-based drag reducers should be replaced by preference for other methods that do not involve polymer additives [22]:

- riblets,
- dimples,
- oscillating walls,
- compliant surfaces,
- microbubbles.

Drag reduction in turbulent flow of water by air bubbles and other additives was assessed [23]. The friction factor versus Reynolds number data exhibited different trends at low and high Reynolds numbers. At high Reynolds number, the two-phase mixture behaves as a pseudohomogeneous system and the friction factor follows single-phase behavior. In contrast, at low Reynolds number, the observed friction factor becomes significantly higher than that of single-phase fluids. These trends were interpreted in terms of the average bubble sizes.

6.5.2 Damping of Transmission of Eddies

One of the mechanisms of drag reduction is that transmission of eddies can be damped by the viscoelastic properties of fluids. The transfer process of an isolated eddy in Maxwell fluids with viscoelastic properties was studied, and the expressions describing such phenomena were obtained [24]. The results of the study showed that eddy transmission was damped significantly with an increase of the viscoelastic properties of the fluids.

6.5.3 Viscoelastic Fluid Thread

In the extensive literature on polymer drag reduction, it has occasionally been reported that a continuous thread of a high-concentration polymer solution injected into the axis of a pipe produces a drag reduction effect on the water flow in the pipe [25]. The thread seems to persist through the length of the pipe and little, if any, diffusion of polymer to the walls of the pipe is apparent.

A polymer of the polyacrylamide type was injected as a 0.5% solution from an axially placed nozzle at the bellmouth entrance. The experiments showed that the central thread provided drag reduction almost equivalent to that of premixed solutions of the same total polymer concentration flowing in the pipe. Overall concentrations of 1–20 ppm were used.

Moreover, the effects were additive: 2-ppm thread overall concentration plus 2 ppm added gave drag reductions equivalent to 4 ppm. Reynolds numbers of up to 300,000 were investigated. In other experiments, a number of different polymer fluids were injected on the centerline of a water pipe flow facility [26]. Two distinct flow regions were identified:

1. Reynolds numbers above 25,000, at which centerline injection acted as a rather efficient mixing device for water-soluble polymer and no drag reduction resulted from materials insoluble in water.
2. Reynolds numbers from 10,000 to 25,000, at which there is strong evidence that under certain conditions a viscoelastic fluid thread can interact with turbulence eddies and reduce the overall flow friction in the pipe.

6.5.4 Polymer Degradation in Turbulent Flow

Drag reduction in turbulent flow is of great potential benefit to many industrial processes, including long-distance transportation of liquids, oil well operations, and transportation of suspensions and slurries, but it is complicated by the problem of polymer degradation. A capillary rheometer was used to investigate the effect of various parameters on polymer degradation in turbulent flow [27].

These parameters included polymer concentration, contraction ratio, pipe length, pipe diameter, number of passes, solvent weight, and molecular weight of the polymer. A commercial organic drag reducer, two grades of polyacrylamide, and a high molecular weight polyisobutylene were used.

In turbulent flow, the polymer degraded more in a poor solvent at low Reynolds numbers, whereas the opposite effect was observed at high Reynolds numbers. The critical Reynolds number, Re_c, or critical apparent shear extensional rate, V/d, was found to increase with polymer concentration and molecular weight as represented by the dimensionless concentration $c(\eta)$.

Polysaccharide guar gum is used as a turbulent drag reducer in aqueous systems. It reduces the friction drag tremendously under a turbulent flow even in small amounts. A study on the mechanical degradation of guar gum has been presented. The effectiveness of drag reduction was measured with a rotating disk apparatus as a function of time.

Two different degradation models were examined: a single-relaxation process and a stretched-exponential model. The stretched-exponential model seems to fit the experimental data better [28].

6.5.5 Drag Reduction in Two-Phase Flow

The drag-reducing properties of a polyacrylamide were tested in air-water two-phase flow with use of a horizontal pipe of 31-mm diameter [29]. The properties of the polymer were tested in single-phase water flow, and the results were found to comply with the reduction in pressure drop found by other workers. Positive effects in two-phase flow were found to depend on the Reynolds number of the liquid flow.

Thus the drag reduction in stratified flow was small or negative. In slug flow, the drag reduction seems to occur in the liquid slug not in the layer below the bubble. The flow regime seems to be unaffected by the polymer. It has been established that in multiphase flow drag reducers act significantly as corrosion inhibitors because they smooth the flow profile near the walls [30].

The effect of polyacrylamide on the heat transfer and the frictional pressure drop behavior of an upward air-water two-phase flow has been investigated [31]. An aqueous polyacrylamide solution with a concentration of 300 ppm was used in the experiments.

The results indicate that the air-water two-phase frictional pressure drops can be reduced from 26.2% to 42.7%. The two-phase heat transfer coefficients can be reduced from 39.7% to 80.8% by the addition of polyacrylamide.

One can explain the pressure drop reduction in the gas-liquid two-phase regime by combining the liquid-phase drag reduction in the liquid layer and liquid slug and the gas-liquid interfacial friction reduction. The heat transfer reduction may be explained according to the liquid-phase heat transfer reduction in the liquid layer and liquid slug due to the reduction or mitigation of the turbulence [31].

6.5.6 Drag Reduction in Gas Flow

Coating

Drag reduction in gas transmission lines can be achieved by application of a thin chemical coating on the pipe's inner surface to reduce friction between the flowing gas and the pipe wall.

An important criterion for gas drag reduction is that the additive can bond strongly to metal surfaces and form a smooth film to mitigate the turbulence at the gas-solid interface [32]. Effective gas drag reducers have properties similar to those of corrosion inhibitors, lubricants, and surfactants. Fatty acid amines or amides exhibit these properties.

Ammonia

For storage or pipeline transportation of natural gas at pressures above 5.5 MPa (800 psi), it is advantageous to add ammonia to the natural gas. The ammonia should not create a liquid phase at the temperature and pressure used. Such a gas mixture of ammonia and natural gas can be compressed or pumped with a lower energy expenditure than would be needed for an equivalent volume of natural gas alone. When more than 4% by volume of ammonia is present, pumping through pipelines is also aided by the refrigerant effect of the ammonia, which reduces the temperature of the gas being transported [33].

6.5.7 Microfibrils

Friction loss in liquids can be reduced with maximum stability by addition of a predetermined amount of selected organopolymeric microfibrils to a liquid [34]. These microfibrils are insoluble but highly dispersible in the liquid.

An organopolymeric microfibril is a solid organic polymer in the form of microfibrils having an average diameter in the range of 100–1000 Å, an average length in the range of 1–500 μm, and an aspect ratio (length/diameter) of 10–1,000,000. Polymeric materials to be processed into microfibrils should be insoluble but highly dispersible in a given liquid.

6.5.8 Drag-Reducing Surfactant Solutions

The behavior of two types of drag-reducing surfactant solutions was studied in turbulent flows in pipes of different diameters [35]. The surfactant systems contained rodlike micelles consisting of equimolar mixtures of *n*-tetradecyltrimethylammonium bromide, *n*-hexadecyltrimethylammonium bromide, and sodium salicylate.

The structure of the turbulence was studied with a laser Doppler velocitymeter in a 50-mm pipe. In the regimes of turbulent flow, both surfactant solutions exhibited characteristic flow regimes. In the regimes of turbulent flow at low Reynolds numbers, velocity profiles similar to those observed for dilute polymer solutions were found, whereas at maximal drag reduction conditions, more S-shaped profiles that show deviations from a logarithmic profile occurred.

6.5.9 Soapy Industrial Cleaner

Experiments have been conducted to investigate the effect of a soapy industrial cleaner on reducing the skin friction of a Jordanian crude oil flowing turbulently in pilot-scale pipes of different sizes. Experiments showed that a concentration of only 2 ppm of the chemical additive injected into the crude oil line caused an appreciable amount of drag reduction [36]. The effects of additive concentration and pipe diameter on drag reduction have been investigated.

6.5.10 Lyophobic Performance of the Lining Material

An experimental study was conducted on the characteristics of frictional drag for a lyophobic surface, with the surface consisting of Teflon and the working media being water and machine oil [29]. The test results indicate that, depending on the lyophobic performance of the lining material, the pipes lined with Teflon had a better drag-reducing effect than conventional steel pipes.

A drag reduction of approximately 12% was achieved with water as the working medium or 6% with machine oil as the working medium. In other words, Teflon has a higher lyophobic performance against water than against machine oil.

The theoretical analysis of the flow mechanism on the lyophobic surface shows that the lyophobic treatment of the surface can lower the surface energy to such a degree that the attraction of liquid molecules to the solid wall becomes weaker than the liquid molecular absorption. This effect causes a gliding flow adjacent to the pipe wall, thus reducing the drag.

6.5.11 Emulsions

The effects of the concentrations of surfactant and water of emulsions on the flow in pipelines have been investigated [37].

For a fixed water concentration of 30% by volume, the flow behavior of water-in-oil emulsions with different surfactant concentrations was measured. As surfactant the oil-soluble sorbitan trioleate was used. The influence of water concentration on the pipeline flow behavior of emulsions was determined by variation of the water concentration from 30 to 80%. When the water concentration is increased above 45% by volume, an inversion of the water-in-oil emulsion into an oil-in-water emulsion occurs. The water-in-oil emulsion without surfactant is unstable and exhibits a drag reduction behavior in turbulent flow [37].

The fluidity of waxy crude oil with a pour point of 47°C could be greatly improved by emulsification with a synthetic formation water [38]. It was found that the combination of cocamidopropylamine oxide (CAO-35) and sodium oleate is an effective emulsifier mixture to form a stable waxy crude oil-in-water emulsion.

The optimum composition of emulsifying additives with respect to the total mass of the emulsion was obtained as given in Table 6.2.

It was found that the pour point of the emulsion increased as the oil content increased, and the optimum mixing speed and cooling rate were 600 rpm and 0.5°C min^{-1}, respectively.

Under the optimum emulsifying conditions, for oil-water emulsions with an oil content of 70%, a pour point reduction of 25°C was observed

Table 6.2 Emulsifying composition

Compound	Content (%)
CAO-35 and sodium oleate (8:2)	0.4
Sodium triphosphate	0.028
NaOH	0.05
Polyacrylamide	0.15

Source: Gao Y, Wu B, Gao J, Wang L, Zhao W, Gui Z. The formation of waxy crude oil-in-water emulsions for the reduction of pour point and viscosity. Petrol Sci Tech 2013;31(23):2443–50. http://dx.doi.org/10.1080/10916466.2011.619242.

and the viscosity reduction was 97.46% at 40°C. So such compositions are highly promising for the exploitation and transportation of waxy crude oil [38].

Herschel-Bulkley Model

The emulsification method has been thoroughly evaluated by investigation of the rheological properties of crude oil-in-water emulsions [39]. The factors affecting the rheology of a heavy crude oil-in-water emulsion stabilized by sodium carbonate as an actuator of natural surfactant (ie, asphaltene) have been studied. An Iranian heavy crude oil with an almost Newtonian rheological behavior (West Paydar) was used as the oil phase. In contrast, the rheological behavior of the emulsions was non-Newtonian. The rheological model of Herschel and Bulkley [40, 41] was fitted to the data of both emulsions and heavy crude. The Herschel-Bulkley model was capable of predicting the pressure loss under each condition investigated and also of optimizing the required and intended process parameters for transportation.

Core Annular Flow

The pressure drop for fluid transportation systems carrying heavy oil can be reduced with the concept of core annular flow [42].

The method involves the formation of a biphasic flow system wherein a higher viscosity fluid is the core and a lower viscosity fluid is injected as a surrounding annulus. The biphasic fluid is introduced into a fluid transportation system, such as a pipeline, and is propagated through the length of the fluid transportation system. For heavy oil transport, the heavy oil is the core and water is the annulus.

Table 6.3 Demulsifier additives

Compound
1-Naphthalenesulfonic acid
2,6-Naphthalenedisulfonic acid
1,5-Naphthalenedisulfonic acid
1,3,6-Naphthalenetrisulfonic acid
1,3,6,8-Pyrenetetrasulfonic acid

Source: Varadaraj R, Thomas ER. Core annular flow of heavy crude oils in transportation pipelines and production wellbores. US Patent 8 298 997, assigned to ExxonMobil Upstream Research Company (Houston, TX); 2012. http://www.freepatentsonline.com/8298997.html.

Core annular flow of heavy oil has been examined; however, core annular flows in fluid transportation systems have not been widely practiced. One obstacle is that conventional tubing and pipeline conduits have an affinity for heavy oil, resulting in adhesion of heavy oil.

The reduction of the friction within the pipeline flow regime has been achieved by polymeric drag reducers or an oleophobic film-forming agent in the water phase.

More recently, demulsifier additives have been described for the separation of fluids in a biphasic flow system. These additives are sodium salts of polynuclear aromatic sulfonic acids. Examples of such additives are summarized in Table 6.3.

6.5.12 Interpolymer Complexes

It has been shown that hydrogen bonding-mediated interpolymer complexes can be powerful drag reducers. The drag reduction levels in such polymer systems increase dramatically by a factor of 2–6 when compared with their nonassociating polymeric precursors. Their shear stability is also shown to be significantly enhanced [43].

Hydrocarbon soluble polymers containing low percentages of polar associating groups are used to determine the effects of polymer associations on solution drag reduction. Experimental data suggest that intrapolymer associations generally decrease the dilute solution drag reduction activities of single associating polymers with like polar groups [44]. Interpolymer complexes formed by one polymer with anionic groups and one polymer with cationic groups can overcome this limitation and provide an enhanced dilute solution drag reduction activity as a result of favorable interpolymer associations, which build larger structures of higher apparent molecular weight. The latter associations may also increase the resistance of the polymers to degradation in turbulent flows.

6.6 DRAG-REDUCER CHEMICALS

Drag reducer types are summarized in Table 6.4 and some of them are discussed in detail below.

6.6.1 Chelating Agents

Amine-chelate complexes can be added to heavy crude oils to reduce their viscosity significantly [52]. Amine-chelate complexes can be synthesized by the heating of an organic amine with a chelating agent. Some tertiary alkyl primary amines are summarized in Table 6.5.

Table 6.4 Flow improvers

Compound	Reference
Hexatriethanolamine monooleate	[45]
Ultra high molecular weight polyethylene	[46]
Esters of a oleic acid–maleic anhydride copolymer	[47]
Copolymers from styrene and maleic anhydride with octadecyl amine	[48]
Copolymers from styrene and maleic anhydride with octadecyl amine	[49]
Alkyl-substituted phenol formaldehyde resin and a solvent	[50]
Copolymers from vinyl acetate and methacrylic acid esterified with stearyl and behenyl alcohols	[51]

Table 6.5 Amines for chelate complexes

Amine	
n-Octylamine	2-Ethylhexylamine
tert-Octylamine	*n*-Decylamine
Tridecylamine	*n*-Undecylamine
Lauryl amine	Hexadecylamine
Heptadecylamine	Octadecylamine
Decenylamine	Dodecenylamine
Palmitoleylamine	Oleylamine
Linoleylamine	Eicosenylamine

Source: Chheda BD, Banavali RM, Mazza G. Recovery and transportation of heavy crude oils. US Patent 6 402 934, assigned to Rohm and Haas Company (Philadelphia, PA); 2002. http://www.freepatentsonline.com/6402934.html.

Table 6.6 Viscosity reduction by amino chelates

Temperature (°C)	Additive (ppm)	Viscosity (cP)	Reduction (%)
24	0	6580	–
	750	5650	17.5
	1000	5450	20.4
	1250	5200	24.1
	1750	4500	34.3
32	0	2865	–
	750	2312	19.3
	1000	2125	25.8
	1250	1875	34.6
	1750	1462	49.0
63	0	492	–
	750	440	10.6
	1000	420	14.6
	1250	390	20.7
	1750	340	30.9

Source: Chheda BD, Banavali RM, Mazza G. Recovery and transportation of heavy crude oils. US Patent 6 402 934, assigned to Rohm and Haas Company (Philadelphia, PA); 2002. http://www.freepatentsonline.com/6402934.html.

In particular, a mixture of C_{12} to C_{14} tertiary alkyl primary amines has been heated with ethylenediaminetetraacetic acid or nitrilotriacetic acid as the chelating agent. The synthesis has been exemplified in detail [52].

The effect of both the concentration of the complex synthesized as explained above and the temperature on the viscosity of heavy crude oil for use in a pipeline is shown in Table 6.6.

6.6.2 Ultra High Molecular Weight Polyethylene

The flow of liquid hydrocarbons can be enhanced by introduction into the stream of a nonagglomerating suspension of ultra high molecular weight polyethylene [46, 53] in water with small amounts of surfactant. The finely divided ultra high molecular weight polyethylene is prepared by polymerization and then cryoground below the glass transition temperature.

6.6.3 Copolymers of α-Olefins

Several copolymers of α-olefins are used as drag reducers. Suggested recipes are summarized in Table 6.7. Monomers are shown in Fig. 6.2.

Table 6.7 Copolymers from α-olefins and compounds used as drag reducers

Comonomer	References
Divinylbenzene/1-hexene, 1-octene, 1-decene, and 1-dodecene[a]	[8, 54]
Styrene/*N*-vinyl-2-pyrrolidone	[55]
Ethene/α-olefins[b]	[56]
Homopolymers or copolymers α-olefins[c]	[57]
Polyisobutene[d]	[58]
Methacrylic acid esters	[14]
C_{12}-C_{18} acrylate or methacrylate/ionic monomer[e]	[13]
tert-Butylstyrene/alkyl acrylate, methacrylic acid	[15]
Acrylic acid acrylate	[59]
Ultra high molecular weight polyolefin	[60]
Styrene/methyl styrene sulfonate/*N*-vinylpyridine[f]	[61]

[a]Synthesis by a Ziegler-Natta process.
[b]Up to C_{30}, Ziegler-Natta process.
[c]Molecular weight of up to 15,000 and an isotacticity of 75% or greater.
[d]Oil-soluble polymer.
[e]Reduces friction in the flow of a hydrocarbon fluid by a factor of 5 at concentrations as low as 1–25 ppm.
[f]Polyampholytes.

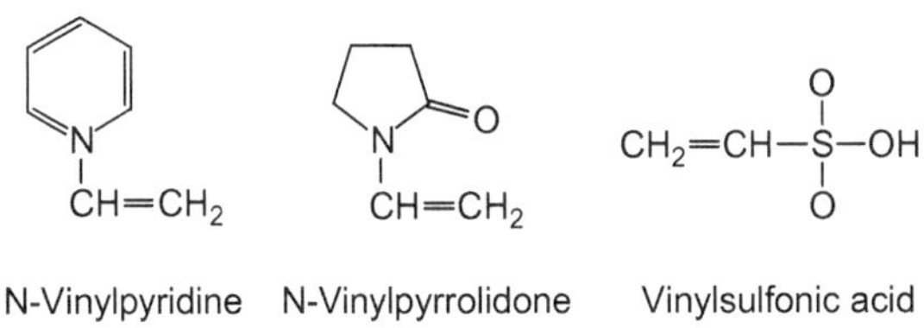

Figure 6.2 Vinyl monomers.

Linear low density polyethylene is a copolymer of ethylene and α-olefins. It is obtained by copolymerization with use of Ziegler-Natta catalysts or metallocene catalysts. Concentrates may be prepared by precipitation of the polymer from a kerosene solution with 2-propanol [62]. The resulting slurry concentrate dissolves rapidly in flowing hydrocarbon streams.

By coating poly(α-olefin) with a fatty acid wax as a partitioning agent and dispersing it in a long-chain alcohol, one can obtain a nonagglomerating, nonaqueous suspension [63].

6.6.4 Maleic Anhydride Copolymers

Polymeric flow improvers belong to the most important chemicals used in the oil industry, in particular in the context of waxy crudes [48].

A series of flow improvers were synthesized by the reaction between copolymers from styrene and maleic anhydride with octadecylamine [49]. The materials had different degrees of imidization; thus the ratio of the maleamic acid group and the maleimide group can be tailored.

The flow improvers were applied to a model waxy oil with C_{24} as the paraffin. Experiments indicated that addition of only a small amount of the flow improver can significantly reduce the crystallization temperature, the number and size of waxy crystals, and the yield stress of the model oil.

In addition, the degree of imidization has a considerable effect on the performance, Flow improvers with a higher degree of imidization are more effective. This effect is believed to occur because of the difference in the polarity of the maleamic acid group and the maleimide group [49].

A family of additives have been prepared by the chemical modification of copolymers from octadecene and maleic anhydride [48]. After polymerization, the copolymers were chemically modified by the grafting of alcohols, such as dodecanol, hexadecanol, or docosanol. In this way, comblike copolymers were obtained. These copolymers were investigated as flow improvers. The performance as a pour point depressant and as a viscosity improver was characterized by the rheological behavior. The additive must exhibit a certain hydrophobic-lypophobic balance and low crystallinity to be effective in inhibiting depositions of paraffins [48].

Similar comblike copolymers were use to improve the flowability of Liaohe extra heavy oil [64]. This oil has high paraffin and asphaltene contents and a pour point of up to 60°C.

Model oils with paraffin mixtures with the same average carbon number as Liaohe extra heavy oil with and without asphaltenes were prepared to explore the effect of the polymers on paraffin crystallization and asphaltene dispersion. It was found that poly(maleic alkylamide-*co*-α-octadecene) copolymers reduce the yield stress and change both the size and the shape of the deposited paraffin crystals. Also, they may obstruct the paraffin crystallization.

6.6.5 Vinyl Acetate and Acrylic Acid Copolymers

Copolymers based on vinyl acetate and methacrylic acid monomers have been prepared with different monomer ratios. The copolymers were esterified with stearyl and behenyl alcohols [51]. The efficiency of the copolymers prepared as pour point depressants and flow improvers was studied. It was found that the ester compounds were efficient as both pour point depressants and flow improvers for the crude oil tested [51].

Copolymers of ethylene acrylic acid were esterified with 1-docosonol to obtain an ester copolymer and this was then grafted with vinyl acetate [65]. The efficiency of these copolymers as pour point depressants is quite good as there is a reduction of the pour point temperature from 27 to 6°C, depending on the composition and the location of the crude oil tested.

6.6.6 Oleic Acid–Maleic Anhydride Copolymer

Oleic acid–maleic anhydride copolymer was esterified by long-chain fatty alcohols to obtain flow improvers [47]. The maximum depression of the pour point was obtained by use of a long-chain alcohol with 22 carbon atoms. In contrast, no depression was obtained with an alcohol with 18 carbon atoms. Further, it was found that the alcohol with 22 carbon atoms enhanced the Bingham yield values. Photomicrography analysis indicated that the wax morphology was modified to fine dispersed crystals of compact size [47].

6.6.7 Phthalimide and Succinimide Copolymers

The synthesis and characterization of phthalimide and succinimide copolymers with vinyl acetate, styrene, and methyl methacrylate have been described. The materials were evaluated for their performance as flow improvers [66].

N-Hydroxymethyl phthalimide was synthesized by the reaction of phthalic anhydride with urea and then with formaldehyde as shown in Fig. 6.3. Eventually, the *N*-hydroxymethyl phthalimide was esterified with acrylic acid to obtain *N*-methylacrylate phthalimide.

In a similar way, the related succinimide monomer was synthesized. These monomers were used as comonomers for the polymerization with styrene, vinyl acetate, and methyl methacrylate.

The molecular weights, polydispersity index, and nitrogen content of the copolymers drastically affect their efficiency as pour point depressants and as flow improvers. With decreasing concentration of the additives, an increase of their activity was obtained. Further, it was found that among the polymers studied, a styrene phthalimide copolymer exhibited the maximum pour point depression [66].

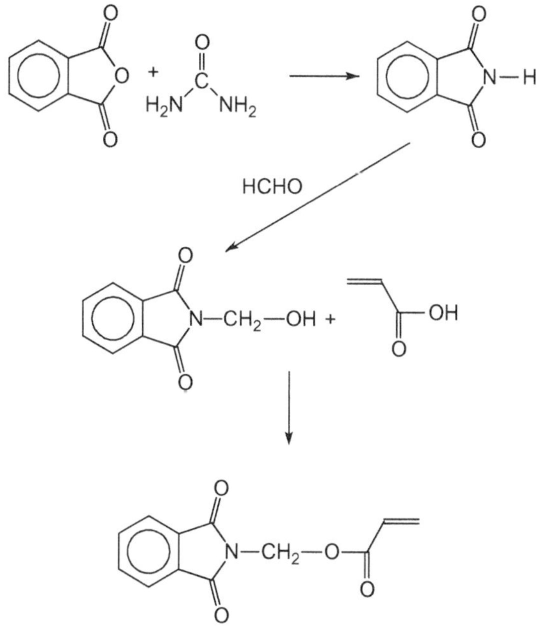

Figure 6.3 Synthesis of *N*-methylacrylate phthalimide. *(Source: Al-Sabagh AM, El-Hamouly SH, Khidr TT, El-Ghazawy RA, HigazySA. Preparation the esters of oleic acid-maleic anhydride copolymer and their evaluation as flow improvers for waxy crude oil. J Dispers Sci Tech 2013;34(11):158596. http://dx.doi.org/10.1080/01932691.2012.739857).*

6.6.8 Phenol Formaldehyde Resin

A drag-reducing additive for heavy oil, such as crude oil, includes an alkyl-substituted phenol formaldehyde resin and a solvent. The phenol formaldehyde resin may have a weight-average molecular weight of 2000–7000.

As solvents, a mixture of ethyl acetate, butyraldehyde, toluene, and 2-propanol can be used. Compared with an otherwise identical heavy oil without the additive, this additive may reduce the viscosity of the combined oil, diluent, and additive by at least 20%, increase throughput by at least 6%, reduce power consumption by at least 3%, reduce the diluent proportion by at least 3%, or cause some combination of these effects [50, 67]. In Table 6.8 the effects of some formulations are shown.

6.6.9 Polyether Compounds for Oil-Based Well Drilling Fluids

A liquid oil, an emulsifier, and a friction modifier, which includes certain polyether compounds, can be added to a drilling fluid consisting of a

Table 6.8 Viscosity reduction

Ingredient	**Amount (%)**				
Formaldehyde resin	5	5	5	5	5
Solvent mixture					
Ethyl acetate	–	47.5	–	95	85
Butyraldehyde	47.5	–	–	–	–
Toluene	47.5	47.5	95	–	–
2-Propanol	–	–	–	–	10
Viscosity reduction (%)	18	20	21	22	23
Improved flow (%)	2	1.9	3	2.7	2.1

Source: Asomaning S, Lehrer SE. Additive to improve flow, reduce power consumption and pressure drop in heavy oil pipelines. US Patent 8 342 198, assigned to Baker Hughes Incorporated (Houston, TX); 2013. http://www.freepatentsonline.com/8342198.html.

water-in-oil emulsion formed from a brine [68]. The friction modifier serves to decrease the coefficient of friction of the well drilling fluid.

Decreasing the coefficient of friction lowers the force required to turn the drill bit in the hole. Gravitational forces increase the coefficient of friction in deviated, horizontal, and extended-reach wells.

6.6.10 Poly(vinyl alcohol)

Heavy crude oil that is too viscous to enable it to be pumped from a flowing phase of a reservoir into and along a pipeline for delivery to a refinery may be contacted with a formulation containing poly(vinyl alcohol) in order to reduce its viscosity [69]. Also a cross-linked poly(vinyl alcohol) may be used for a subsequent grafting.

Poly(vinyl acetate) which has been 60–95% hydrolyzed to poly(vinyl alcohol) is used. In addition, poly(1,4-di(4-(*N*-methylpyridinyl))-2,3-di(4-(1-formylphenyl)butylidene) may be added as a second polymer. One can prepare this polymer from an aqueous solution of 4-(4-formylphenylethenyl)-1-methylpyridinium methosulfonate (see Fig. 6.4) by exposing it to ultraviolet light. This results in a photochemical reaction between the carbon-carbon double bonds of adjacent 4-(4-formylphenylethenyl)-1-methylpyridinium methosulfate molecules, thus producing the desired polymer [69, 70]. In a further reaction, the pendent formyl groups may be reacted with poly(vinyl alcohol) to produce a grafted polymer.

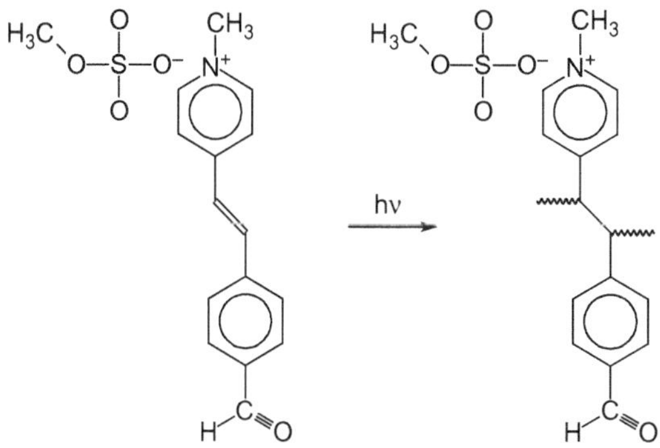

Figure 6.4 4-(4-Formylphenylethenyl)-1-methylpyridinium methosulfonate.

The photosensitivity of the styrylpyridinium group was first reported in 1977 [71].

The properties of styrylpyridinium salts in aqueous solution have been described as somewhat unusual, because agglomerates are formed [72]. After the viscous composition has been transported to a desired location, it may be separated from the other components.

6.6.11 Latex Drag Reducers

Latex drag reducers comprise a polymer that is formed via an emulsion polymerization reaction dispersed in a continuous phase and can undergo subsequent modification in order to increase the solubility of the polymer in hydrocarbons.

The monomer is preferably 2-ethylhexyl methacrylate. The polymerization is done by conventional emulsion polymerization techniques. Details of the preparation procedure can be found elsewhere [2]. The emulsion polymerization reaction yields an initial latex composition, which is a stable colloidal dispersion.

The dispersed phase is made up of colloidal particles of the high molecular weight polymer, with totally approximately 50% by weight of the initial latex. The continuous phase is water and surfactant. The latex can be modified or formulated with additional surfactant and organic solvents. In this way, the viscosity may be decreased.

A schematic diagram of an engineering loop recirculation test apparatus used to measure the effectiveness of drag reducers is shown in Fig. 6.5. This test allows the evaluation of drag reducer performance when the drag

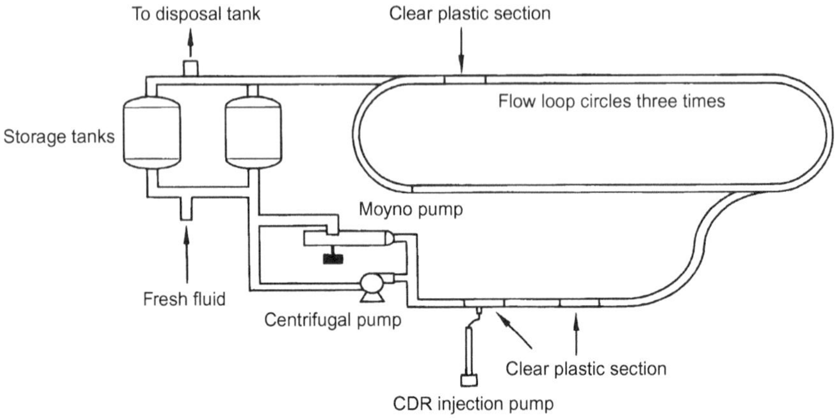

Figure 6.5 Loop recirculation test apparatus. *(Source: Milligan SN, Harris WF, Smith KW, Burden TL, Johnston RL, AndersonVS. Remote delivery of latex drag-reducing agent without introduction of immiscible low-viscosity flow facilitator. US Patent 7 361 628, assigned to ConocoPhillips Company (Houston, TX); 2008. http://www.freepatentsonline.com/7361628.html).*

Table 6.9 High hydrophilic-lipophilic balance anionic and nonionic surfactants

Anionic surfactants	Nonionic surfactants
Sodium lauryl sulfate	Sorbitan esters
Dioctyl sodium sulfosuccinate	Ethoxylated glycerine esters
2-Ethylhexyl polyphosphate sodium salt	Ethoxylated fatty amines
Sodium dodecylbenzenesulfonate	Fatty acid esters
Sodium lauroyl sarcosinate	Ethoxylated alcohols

Source: Burden TL. Drag reduction of asphaltenic crude oils. US Patent 8 656 950, assigned to Phillips 66 Company (Houston, TX); 2014. http://www.freepatentsonline.com/8656950.html

reducer is injected in non-predissolved form into a hydrocarbon fluid in the flow loop. The test can be used to simulate the performance profiles and drag reducer behavior in field pipelines over a 3-h period in terms of dissolution, peak performance, and degradation of the drag-reducing polymer.

The drag reducer may be injected into the pipeline by conventional or umbilical delivery systems. The surfactant used should have a high hydrophilic-lipophilic balance number. Exemplary high hydrophilic-lipophilic balance anionic and nonionic surfactants are shown in Table 6.9.

6.6.12 Biopolymers

Several biopolymers have been found that exhibit drag-reducing capabilities in aqueous media. Thus, biopolymers would make a suitable alternative to artificial drag reducers. The state of the art of biopolymers as drag reducers and new possibilities have been reviewed [73].

Carboxymethyl Cellulose

Carboxymethyl cellulose is a synthetic polymer which will seep into the ground and further harm our environment. A more environmentally friendly drag-reducing agent, such as a polymer derived from natural sources or a biopolymer, is required to avoid seeping [74].

As opposed to synthetic polymers, the potential of biopolymers as drag reduction agents, especially those derived from a local plant source, has not been extensively explored. The drag reduction of a polymer produced from a local plant source within the turbulent regime have been explored and assessed with a rheometer, where a reduced torque produced can be perceived as a reduction of drag.

Tylose

Tylose is not as effective in drag reduction as other substances described in the literature. Detailed mean velocity, normal Reynolds stress, and pressure drop measurements were performed with 0.4–0.6% aqueous solutions of Tylose, a methylhydroxyl cellulose type (molecular weight 6000), after a selection process from a set of low molecular weight fluids [35].

The measurements of the viscosity of the Tylose solutions show shear thinning behavior, and the oscillatory and creep tests measured elastic components of the stress on the order of the minimal detectable values by the rheometer. These low molecular weight polymer solutions delay the transition from the laminar to the turbulent regime and show drag reductions of half those that reportedly occur with other low-elasticity, shear-thinning, high molecular weight, aqueous polymer solutions.

6.6.13 Microencapsulated Polymers

Highly concentrated drag-reducing agents may be prepared by microencapsulation of a polymer or a monomer. The microencapsulation may be performed before, during, or after the polymerization of a monomer to an effective drag-reducing polymer. If the encapsulation is done before or during polymerization, a catalyst must be present, but little or no solvent is required. The result is a bulk polymerization within the microcapsule.

The inert capsule or shell may be removed before, during, or after the introduction of the microencapsulated drag reducer into a flowing liquid. No injection probes or other special equipment should be required to introduce the drag-reducing slurry into the liquid stream, nor is grinding (cryogenic or otherwise) of the polymer necessary to form a suitable drag-reducing agent [75].

6.6.14 Aluminum Carboxylate

Aluminum carboxylate-based drag reducers are nonpolymeric drag-reducing agents. These drag-reducing additives are not subject to shear degradation and do not cause undesirable changes in the emulsion or fluid quality of the fluid being treated, or undesirable foaming.

The compositions consists of an aluminum carboxylate itself and fatty acids. The aluminum carboxylates are selected from aluminum salts of fatty acids, including octoates, stearates, oleates, or naphthenates [76]. The fatty acids are selected from long-chain carboxylic acids. Aluminum salts of a combination of short- and long-chain carboxylic acids may provide an optimum balance between drag reduction and change in viscosity.

REFERENCES

[1] Almond NE. Pipeline flow improvers. In: Proceedings Volume. API Pipeline Conf. (Dallas, 4/17–18/89); 1989. p. 307–11.

[2] Milligan SN, Harris WF, Smith KW, Burden TL, Johnston RL, Anderson VS. Remote delivery of latex drag-reducing agent without introduction of immiscible low-viscosity flow facilitator. US Patent 7 361 628, assigned to ConocoPhillips Company (Houston, TX); 2008. http://www.freepatentsonline.com/7361628.html.

[3] Goudy CFL. How flow improvers can reduce liquid line operating costs. Pipe Line Gas Ind 1991;74(6):49–51.

[4] Muth CL, Kolby SM. Cost saving by use of flow improver. In: Proceedings Volume. 13th Int. Pipeline Technol. Conf. (Houston, 2/5–7/85); 1985. p. 353–7.

[5] Karami HR, Mowla D, Investigation of the effects of various parameters on pressure drop reduction in crude oil pipelines by drag reducing agents. J Non-Newt Fluid Mech 2012;177–178:37–45. http://www.sciencedirect.com/science/article/pii/S0377025712000687, http://dx.doi.org/10.1016/j.jnnfm.2012.04.001.

[6] Karami HR, Mowla D. A general model for predicting drag reduction in crude oil pipelines. J Petrol Sci Eng 2013;111(0):78–86. http://www.sciencedirect.com/science/article/pii/S0920410513002295, http://dx.doi.org/10.1016/j.petrol.2013.08.041.

[7] Aubanel ML, Bailly JC. Amorphous high molecular weight copolymers of ethylene and alpha-olefins. US Patent 4 797 461, assigned to BP Chemicals Limited (London, GB2); 1989. http://www.freepatentsonline.com/4797461.html.

[8] Gessell DE, Washecheck PH. Composition and method for friction loss reduction. US Patent 4 952 738, assigned to Conoco Inc. (Ponca City, OK); 1990. http://www.freepatentsonline.com/4952738.html.

[9] Grabois RN, Lee YN. Use of a water soluble drag reducer in a water/oil/gas system. US Patent 5 027 843, assigned to Conoco Inc. (Ponca City, OK); 1991. http://www.freepatentsonline.com/5027843.html.
[10] Lee YN, Wiggins FR. Activation of water-in-oil emulsions of friction reducing polymers for use in saline fluids. US Patent 5 067 508, assigned to Conoco Inc. (Ponca City, OK); 1991. http://www.freepatentsonline.com/5067508.html.
[11] Majumdar S, Holay SH, Singh RP. Adiabatic compressibility and solvation of drag reducing polymers in aqueous solutions. Euro Polym J 1980;16(12):1201–6.
[12] Karydas A. Use of organic fluorochemical compounds with oleophobic and hydrophobic groups in asphaltenic crude oils as viscosity reducing agents. US Patent 4 876 018, assigned to Ciba-Geigy Corporation (Ardsley, NY); 1989. http://www.freepatentsonline.com/4876018.html.
[13] Malik S, Shintre SN, Mashelkar RA. Process for the preparation of a new polymer useful for drag reduction in hydrocarbon fluids in exceptionally dilute polymer solutions. US Patent 5 080 121, assigned to Council of Scientific & Industrial Research (New Delhi, IN); 1992. http://www.freepatentsonline.com/5080121.html.
[14] Ritter W, Meyer C, Zoellner W, Herold CP, Tapavicza SV. Copolymers of (meth) acrylic acid esters as flow improvers in oils. US Patent 5 039 432, assigned to Henkel Kommanditgesellschaft auf Aktien (Duesseldorf-Holthausen, DE), 1991. http://www.freepatentsonline.com/5039432.html.
[15] Naiman MI, Chang JC. Methods and compositions for reduction of drag in hydrocarbon fluids. US Patent 4 983 186, assigned to Petrolite Corporation (St. Louis, MO); 1991. http://www.freepatentsonline.com/4983186.html.
[16] Oh-Kil K, Ling-Siu C. Drag reducing polymers. In: The polymeric materials encyclopedia. Boca Raton, FL: CRC Press, Inc.; 1996.
[17] Al-Sarkhi A. Drag reduction with polymers in gas-liquid/liquid-liquid flows in pipes: a literature review. J Nat Gas Sci Eng 2010;2(1):41–8. http://www.sciencedirect.com/science/article/B94T1-4YG1W36-1/2/dbc30103cfe9d5279f1eaba7b54cdf60.
[18] Gampert B, Wagner P. The influence of polymer additives on velocity and temperature fluid. Berlin, Germany: Springer-Verlag; 1985. p. 71.
[19] Chang HFD, Meng JS, Prediction of drag reduction from rheological properties of dilute polymer solutions. Physicochem Hydrodyn 1987;9:33.
[20] Deshmukh SR, Chaturved PN, Singh RP. Novel biodegradable flocculants based on polysaccharides. J Appl Polym Sci 1985;30:4013–20.
[21] Kostic M. On turbulent drag and heat transfer reduction phenomena and laminar heat transfer enhancement in non-circular duct flow of certain non-newtonian fluids. Int J Heat Mass Transfer 1994;37:133–47.
[22] Abdulbari HA, Yunus RM, Abdurahman NH, Charles A, Going against the flow: a review of non-additive means of drag reduction. J Ind Eng Chem 2013;19(1):27–36. http://www.sciencedirect.com/science/article/pii/S1226086X1200247X, http://dx.doi.org/10.1016/j.jiec.2012.07.023.
[23] Baghaei P, Pal R. Effect of bubbles and additives on friction factor in pipeline flow. Can J Chem Eng 2013;92(3):543–52. http://dx.doi.org/10.1002/cjce.21858.
[24] Li Z. Effect of fluid viscoelasticity on isolated eddy transmission. J Univ Petrol China 1991;15(5):33–8.
[25] Hoyt JW, Sellin RHJ. Drag reduction by centrally-injected polymer "threads". Rheol Acta 1988;27(5):518–22.
[26] Hoyt JW, Sellin RHJ. Polymer "threads" and drag reduction. Rheol Acta 1991;30(4):307–15.
[27] Moussa T, Tiu C. Factors affecting polymer degradation in turbulent pipe flow. Chem Eng Sci 1994;49(10):1681–92.

[28] Hong CH, Zhang K, Choi HJ, Yoon SM. Mechanical degradation of polysaccharide guar gum under turbulent flow. Ind Eng Chem 2010;16(2): 178–80. http://www.sciencedirect.com/science/article/B8JDF-4YDTS2G-D/2/5b3f001ee48947fee5f883608e87849c.
[29] Saether G, Kubberud K, Nuland S, Lingelem M, Drag reduction in two-phase flow. In: Proceedings Volume. 4th Bhra Multi-Phase Flow Int. Conf. (Nice, Fr, 6/19–21/89); 1989. p. 171–84.
[30] Kang C, Jepson WP, Gopal M. The effect of drag reducing agents on corrosion in multiphase flow. In: Proceedings Volume. NACE Int. Corrosion Conf. (Corrosion 98) (San Diego, 3/22–27/98); 1998.
[31] Liu L, Cheng L. Effect of a polymer additive on heat transfer and pressure drop behaviors of upward air–water flow in an inclined smooth circular tube. Heat Transfer Eng 2013;34(13):1099–111. http://dx.doi.org/10.1080/01457632.2013.763549.
[32] Li YH, Chesnut GR, Richmond RD, Beer GL, Caldarera VP. Laboratory tests and field implementation of gas drag reduction chemicals. In: Proceedings Volume. SPE Oilfield Chem. Int. Symp. (Houston, 2/18–21/97); 1997. p. 457–69.
[33] Morris I, Perry G. High pressure storage and transport of natural gas containing added c2 or c3, or ammonia, hydrogen fluoride or carbon monoxide. US Patent 6 217 626, assigned to JL Energy Transportation Inc. (Calgary, CA); 2001. http://www.freepatentsonline.com/6217626.html.
[34] Shinomura T. Method of reducing friction losses in flowing liquids. US Patent 4 751 937, assigned to Nippon Oil Co., Ltd. (Minato, JP); 1988. http://www.freepatentsonline.com/4751937.html.
[35] Bewersdorff HW, Ohlendorf D. The behaviour of drag-reducing cationic surfactant solutions. Colloid Polym Sci 1988;266(10):941–53.
[36] Mansour AR, Aldoss T. Drag reduction in pipes carrying crude oil using an industrial cleaner. SPE Unsolicited Pap 1988; (17918-MS).
[37] Omer A, Pal R. Effects of surfactant and water concentrations on pipeline flow of emulsions. Ind Eng Chem Res 2013;52(26):9099–105. http://dx.doi.org/10.1021/ie3036492.
[38] Gao Y, Wu B, Gao J, Wang L, Zhao W, Gui Z. The formation of waxy crude oil-in-water emulsions for the reduction of pour point and viscosity. Petrol Sci Tech 2013;31(23):2443–50. http://dx.doi.org/10.1080/10916466.2011.619242.
[39] Hoshyargar V, Ashrafizadeh SN. Optimization of flow parameters of heavy crude oil-in-water emulsions through pipelines. Ind Eng Chem Res 2013;52(4):1600–11. http://dx.doi.org/10.1021/ie302993m.
[40] Herschel WH, Bulkley R, Konsistenzmessungen von gummi-benzollsungen. Kolloid-Zeitschrift 1926;39(4):291–300. http://dx.doi.org/10.1007/BF01432034.
[41] Tang HS, Kalyon DM, Estimation of the parameters of Herschel-Bulkley fluid under wall slip using a combination of capillary and squeeze flow viscometers. Rheol Acta 2004;43(1):80–8. http://dx.doi.org/10.1007/s00397-003-0322-y.
[42] Varadaraj R, Thomas ER. Core annular flow of heavy crude oils in transportation pipelines and production wellbores. US Patent 8 298 997, assigned to ExxonMobil Upstream Research Company (Houston, TX); 2012. http://www.freepatentsonline.com/8298997.html.
[43] Malik S, Mashelkar RA. Hydrogen bonding mediated shear stable clusters as drag reducers. Chem Eng Sci 1995;50(1):105–16.
[44] Kowalik RM, Duvdevani I, Lundberg RD, Schulz DN, Peiffer DG, Kitano K. Enhanced drag reduction via interpolymer associations. J Non-Newtonian Fluid Mech 1987;24(1):1–10.
[45] Sharma S, Mahto V, Sharma VP. Effect of flow improvers on rheological and microscopic properties of Indian waxy crude oil. Ind Eng Chem Res 2014;53(12):4525–33. http://dx.doi.org/10.1021/ie403050y.

[46] Dindi A, Johnston RL, Lee YN, Massouda DF. Slurry drag reducer. US Patent 5 539 044, assigned to Conoco In. (DE); 1996. http://www.freepatentsonline.com/5539044.html.
[47] Al-Sabagh AM, El-Hamouly SH, Khidr TT, El-Ghazawy RA, Higazy SA. Preparation the esters of oleic acid-maleic anhydride copolymer and their evaluation as flow improvers for waxy crude oil. J Dispers Sci Tech 2013;34(11):1585–96. http://dx.doi.org/10.1080/01932691.2012.739857.
[48] El-Ghazawy RA, Atta AM, Kabel KI. Modified maleic anhydride-co-octadecene copolymers as flow improver for waxy Egyptian crude oil. J Petrol Sci Eng 2014; (0). http://www.sciencedirect.com/science/article/pii/S0920410514002393, http://dx.doi.org/10.1016/j.petrol.2014.07.040.
[49] Cao K, Wei Xx, Li Bj, Zhang Js, Yao Z. Study of the influence of imidization degree of poly(styrene-*co*-octadecyl maleimide) as waxy crude oil flow improvers. Energy Fuels 2013;27(2):640–5. http://dx.doi.org/10.1021/ef301336n.
[50] Asomaning S, Lehrer SE. Additive to improve flow, reduce power consumption and pressure drop in heavy oil pipelines. US Patent 8 575 082, assigned to Baker Hughes Incorporated (Houston, TX); 2013. http://www.freepatentsonline.com/8575082.html.
[51] Atta AM, El-Ghazawy RA, Morsy FA, Ali AMS, Elmor A. Synthesis and characterization of polymeric additives and their effect on flow properties of waxy Egyptian crude oil. Glob J Sci Front Res 2013;13(3). http://www.journalofscience.org/index.php/GJSFR/article/view/755.
[52] Chheda BD, Banavali RM, Mazza G. Recovery and transportation of heavy crude oils. US Patent 6 402 934, assigned to Rohm and Haas Company (Philadelphia, PA); 2002. http://www.freepatentsonline.com/6402934.html.
[53] Smith KW, Haynes LV, Massouda DF. Solvent free oil soluble drag reducing polymer suspension. US Patent 5 449 732, assigned to Conoco Inc. (Ponca City, OK); 1995. http://www.freepatentsonline.com/5449732.html.
[54] Eaton GB, Monahan MJ. Composition of and process for forming polyalphaolefin drag reducing agents. US Patent 5 869 570, assigned to Energy & Environmental International, L.C. (Houston, TX); 1999. http://www.freepatentsonline.com/5869570.html.
[55] Kowalik RM, Duvdevani I, Kitano K, Schulz DN. Drag reduction agents for hydrocarbon solutions. US Patent 4 625 745, assigned to Exxon Research and Engineering Company (Florham Park, NJ); 1986. http://www.freepatentsonline.com/4625745.html.
[56] Hostetler DE, Kostelnik RJ, Shanti In ZJ. Polymerization process. US Patent 4 845 178, assigned to Pony Industries, Inc. (Malvern, PA); 1989. http://www.freepatentsonline.com/4845178.html.
[57] Rossi A, Chandler JE, Barbour R. Polymers and additive compositions; 1993.
[58] Martischius FD, Raab B, Karau D. Process for improving the drag reducing properties of high-molecular weight polymer solutions in crude oil or refined products (verfahren zur verbesserung der fliesswiderstandsvermindernden eigenschaften hochmolekularer polymerloesungen in rohoel oder raffinerieprodukten); 1990.
[59] Schulz DN, Maurer JJ, Bock J, Kowalik RM. A process for the formation of novel acrylamide acrylate copolymers; 1987.
[60] Mack MP. Improved use of flow improvers. Patent: EP 196350 assigned to Conoco Incorporation, 1986.
[61] Peiffer DG, Kowalik RM, Lundberg RD. Drag reduction with novel hydrocarbon soluble polyampholytes. US Patent 4 640 945, assigned to Exxon Research and Engineering Company (Florham Park, NJ); 1987. http://www.freepatentsonline.com/4640945.html.

[62] Fairchild K, Tipton R, Motier JF, Kommareddi NS. Low viscosity, high concentration drag reducing agent and method therefor. US Patent 5 733 953, assigned to Baker Hughes Incorporated (Houston, TX); 1998. http://www.freepatentsonline.com/5733953.html.
[63] Johnston RL, Lee YN. Nonaqueous drag reducing suspensions. US Patent 6 172 151, assigned to Conoco Inc. (Ponca City, OK); 2001. http://www.freepatentsonline.com/6172151.html.
[64] Jiang H, Xu J, Wei X, Wang T, Wang W, Li L, et al. Flow improvement of liaohe extraheavy oil with comb-type copolymers. J Appl Polym Sci 2014;131(8). http://dx.doi.org/10.1002/app.40082.
[65] Al-Shafy HI, Ismail EA. Studies on the influence of polymeric additives as flow improvers for waxy crude oil. IOSR J Eng 2014;4(7):54–61. http://www.iosrjen.org/Papers/vol4_issue7%20%28part-1%29/I047154614.pdf.
[66] Al-Sabagh AM, El-Hamouly SH, Khidr TT, El-Ghazawy RA, Higazy SA. Synthesis of phthalimide and succinimide copolymers and their evaluation as flow improvers for an Egyptian waxy crude oil. Egypt J Petrol 2013;22(3):381–93. http://www.sciencedirect.com/science/article/pii/S1110062113000779, http://dx.doi.org/10.1016/j.ejpe.2013.10.008.
[67] Asomaning S, Lehrer SE. Additive to improve flow, reduce power consumption and pressure drop in heavy oil pipelines. US Patent 8 342 198, assigned to Baker Hughes Incorporated (Houston, TX); 2013. http://www.freepatentsonline.com/8342198.html.
[68] Malchow Jr GA. Friction modifier for oil-based (invert) well drilling fluids and methods of using the same. US Patent 5 593 953, assigned to The Lubrizol Corporation (Wickliffe, OH); 1997. http://www.freepatentsonline.com/5593953.html.
[69] Fletcher P, Crabtree MJ, Eagland D, Crowther NJ. Method for reducing the viscosity of viscous fluids. US Patent 7 745 500, assigned to Advanced Gel Technology Limited (West Yorkshire, GB); 2010. http://www.freepatentsonline.com/7745500.html.
[70] Eagland D, Crowther NJ. Polymeric material. US Patent 6 660 802, assigned to Advanced Gel Technology Limited (West Yorkshire, GB); 2003. http://www.freepatentsonline.com/6660802.html.
[71] Borden DG, Williams JLR. Makromol Chem 1977;178(11):3035–49. http://dx.doi.org/10.1002/macp.1977.021781108.
[72] Crowther NJ, Eagland D. A styrylpyridinium salt in aqueous solution: unusual solution behaviour. Chem Commun 1997;(1):103–4. http://pubs.rsc.org/En/content/articlepdf/1997/cc/a606605d.
[73] Abdulbari HA, Shabirin A, Abdurrahman HN. Bio-polymers for improving liquid flow in pipelines: a review and future work opportunities. J Ind Eng Chem 2014;20(4):1157–70. http://www.sciencedirect.com/science/article/pii/S1226086X13003614, http://dx.doi.org/10.1016/j.jiec.2013.07.050.
[74] Singh HKG, Jaafar A, Yusup S. The effect of sodium hydroxide on drag reduction using a biopolymer. MATEC Web of Conferences 2014;13:1–5. http://dx.doi.org/10.1051/matecconf/20141302030.
[75] Kommareddi NS, Rzeznik LJ. Microencapsulated drag reducing agents. US Patent 6 126 872, assigned to Baker Hughes Incorporated (Houston, TX); 2000. http://www.freepatentsonline.com/6126872.html.
[76] Jovancicevic V, Campbell S, Ramachandran S, Hammonds P, Weghorn SJ. Aluminum carboxylate drag reducers for hydrocarbon emulsions. US Patent 7 288 506, assigned to Baker Hughes Incorporated (Houston, TX); 2007. http://www.freepatentsonline.com/7288506.html.

CHAPTER 7

Pipeline Cleaning

Even carefully built pipelines have been found to contain up to 10 t of waste materials, mostly iron rust, sand, mud, and welding rods. Even shoes, gloves, and bottles have been removed from pipes. Cleaning can generally be achieved by the pumping of suitable solvents in pigging operations.

7.1 PIGS

Pigs are devices that are propelled by a fluid flow so that they travel through pipelines to remove deposits, displace products, and inspect or maintain the pipeline [1].

Initially, pigs were developed for their use in the hydrocarbon industries. However, pigs have since been adopted for many other many fluid-handling processes.

The different types of pigs, their functions, and their features have been reviewed. The commonest types of pigs are considered as are some of the specialist pigs that have been designed for specific tasks. Also, the recent developments of nonsolid pigs have been described. These developments include gel pigs for use in the hydrocarbon industries and ice pigs for use in the fluid products and process industries [1].

7.2 DEPOSITION OF PARAFFINS

The techniques for paraffin removal and paraffin prevention have been reviewed [2]. In particular, inhibitors for paraffin deposits are copolymers of ethylene with vinyl acetate [3, 4]. These materials lower the pour point of the oil.

During production and transportation of crude oil, crystallization and deposition of paraffin waxes may occur. These issues may cause heavy economic losses to the petroleum industry [5].

Guide to the Practical Use of Chemicals in Refineries and Pipelines
http://dx.doi.org/10.1016/B978-0-12-805412-3.00007-6

The fundamental aspects of these problems have been discussed as have the characterization of the paraffins and their solubility tendencies. *n*-Paraffins are predominantly responsible for these problems.

The effect of certain inhibitors on wax precipitation has been studied by differential scanning calorimetry and cross-polarized microscopy [6]. The results from differential scanning calorimetry for the wax appearance temperature do not show remarkable differences with regard to the presence of the inhibitor. In contrast, cross-polarized microscopy clearly shows differences in both the size and the shape of the deposited paraffin. Obviously, in some cases the inhibitor does not modify the wax appearance temperature but modifies the morphology of the precipitated particles.

In particular, when the environment or bulk temperature is below the cloud point or wax appearance temperature, precipitation of the wax from the bulk fluid may occur [7]. Precipitated wax crystals from crude oil deposits on the inner pipe wall could lead to reduced flow area or even complete blockage. A numerical approach has been presented to estimate the deposition profile and its effect on flow-related parameters. The procedure has been validated with field data from a North Sea offshore production facility. Waxes that may contain up to 80 different paraffin compounds have been reported [8].

The mechanism of crystallization of *n*-paraffins has been detailed [5]. Asphaltenes and other resins have a profound effect on the solubility of *n*-paraffins. Also, the climate of the specific area should be considered for the evaluation of the crystallization potential.

A maleic anhydride copolymer derivative may be used as a paraffin inhibitor. Flocculated asphaltenes provide wax crystallization sites whose presence interferes with the inhibitor. The existence of asphaltene-paraffin complexes was assessed by kinetic dissolution studies [9].

Under favorable conditions, *n*-paraffins form orthorhombic crystals, but unfavorable conditions and the presence of impurities lead to hexagonal or amorphous crystals. Gelation reactions are also affected the same way.

The gelation point of model waxy oils with different wax concentrations in the range of 5–20% was investigated by differential scanning calorimetry. The wax precipitation curves obtained from differential scanning calorimetry indicate that the concentration of the precipitated wax crystals increases gradually with the decrease of the temperature. The gelation point increases gradually with increasing wax concentration as shown in Fig. 7.1.

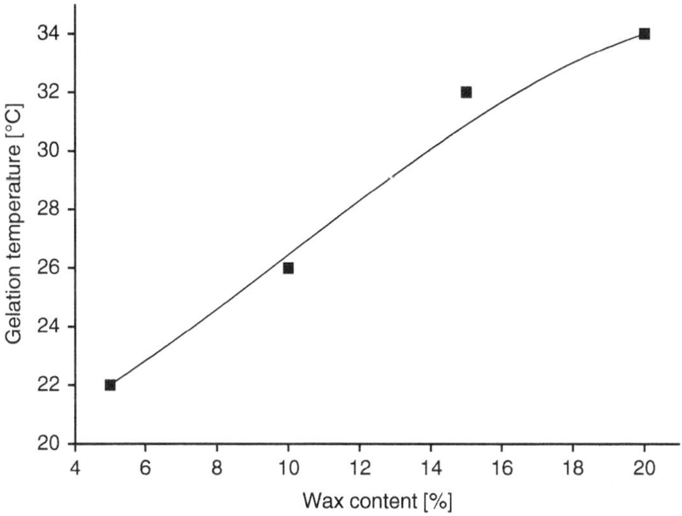

Figure 7.1 Dependence of gelation point on the wax concentration. *(Source: Yang F, Li C, Li C, Wang D. Scaling of structural characteristics of gelled model waxy oils. Energy Fuels 2013;27(7):371824. http://dx.doi.org/10.1021/ef400554v).*

The problems caused by paraffin can be differentiated into those resulting from [5]:

- high pipeline pressure,
- high restarting pressure,
- deposition on the surfaces of the pipe.

7.2.1 Assessment of Deposited Wax

Conventional practices to assess the amount of deposited wax in pipelines are usually based on simulation techniques [10]. Often limited stock tank oil deposition data are collected under laminar-flow conditions in bench-scale flow loops. However, such practices often lead to nonoptimal designs of pipelines and surface facilities.

Improved methods rely on laboratory-scale deposition measurements that are made under realistic conditions. A high-pressure deposition cell that operates on the Taylor-Couette flow principle has been used to generate more deposition data with live reservoir fluids under turbulent flow. This is similar to the conditions encountered in many flow lines [10]. In a case study, scaled-deposition data were obtained and then used to tune a wax deposition model in a simulation package [11].

The deposition of a wax depends on the flow rate, the temperature differential between the crude oil and the pipe surface, the cooling rate, and surface properties [5].

The effect of the surface roughness on the deposition of paraffin was studied. It was concluded that the deposits do not adhere to the metals themselves but are held in place by surface roughness [12]. Further, the deposition of paraffins increases with an increase in surface roughness [13].

It was also observed that temperature has a significant effect on the wax deposition of crude oils [14]. For Indian crude oil, the ideal temperature of the crude oil for pumping could be determined.

7.2.2 Control Methods

One can control paraffin deposition conventionally by remelting the paraffin by circulating hot oil, hot water, or hot air [15]. Also, the use of chemicals such as solvents and dispersants can be successful; however, cost of the chemicals and the pumping equipment required for such a treatment may be high. In addition, since the chemicals usually contain toluene, xylene, methanol, or some other highly toxic, often flammable chemicals, problems may arise.

A microbial treatment, based on the activity of naturally occurring, selectively isolated bacteria, is an effective alternative to conventional methods to prevent and remove paraffin damage [15].

A thermal insulation with aerogel foam had proved to be the most effective option in preventing wax from precipitating [7].

Aerogel

Aerogel is a synthetic porous material derived from a gel in which the liquid component of the gel is replaced with gas. The material has been known since 1931 and has attractive physical properties, such as high temperature resistance, low thermal conductivity, and light weight. Other materials and their minimum insulation thickness required to keep the outlet temperature sufficiently high are summarized in Table 7.1, from which it can be seen that aerogel requires the thinnest insulating layers.

Synergism With Paraffin Deposition

In waxy crude oils, the wax has a tendency to become deposited during storage of the crude oil in tanks or while flowing through pipelines. The deposition in the pipeline severely affects the pipeline throughput. The

Table 7.1 Minimum insulation thickness

Material	Thickness (mm)	Exit temperature (°C)
Polypropylene foam	92	38.4
Polyurethane foam	10	37.9
Microporous material	8	39.4
Aerogel foam	4.7	37.9

Source: Sheikh Mohammad Samiur R, Chacko S. Improved paraffin-deposition profile estimation in hydrocarbon pipelines and effective mitigation review. Oil Gas Facil 2013;2(1):78–85. http://dx.doi.org/10.2118/165933-PA.

deposits have to be removed periodically from the storage tanks and the pipelines by pigging operations.

The wax deposition behaviors of Bombay High crude oil have been studied under different conditions with a cold disk-type assembly [16]. It was found that the deposition occurs much less with additive-treated crude than with untreated crude under otherwise identical conditions. However, the deposits obtained from treated crude have slightly higher melting points and lower oil content. The wax separation temperature of crude oils can be determined from their viscosity behavior [17].

7.3 PIPELINE CLEANING CHEMICALS

The complete evolution of a new cleaning chemical for heavy oil and gas pipelines has been presented [18]. The research involved formulation of the new product and laboratory testing.

Many different methods for cleaning pipeline networks have been developed [19]. Cleaning methods to decrease and remove the iron and iron sulfide deposits include mechanical pigging, batch chemical cleaning, and continuous chemical cleaning. The chemical compounds used in batch and continuous cleaning methods are surfactants, solvents, acids, bases, oxidizing agents, and chelating agents.

Oil naturally contains paraffins, asphaltenes, and naphthenates. During transport, these hydrocarbon components can precipitate and adhere to the pipeline walls and can become associated with iron sulfide scale. Then, corrosion may occur on the pipeline walls below these deposits.

7.3.1 Carbon Disulfide

Wax, paraffin, or sulfur deposits in an oil or gas pipeline can be dissolved by addition of a compound which liberates carbon disulfide on contact with acid. Thereby the deposit is dissolved by the liberated carbon disulfide [20].

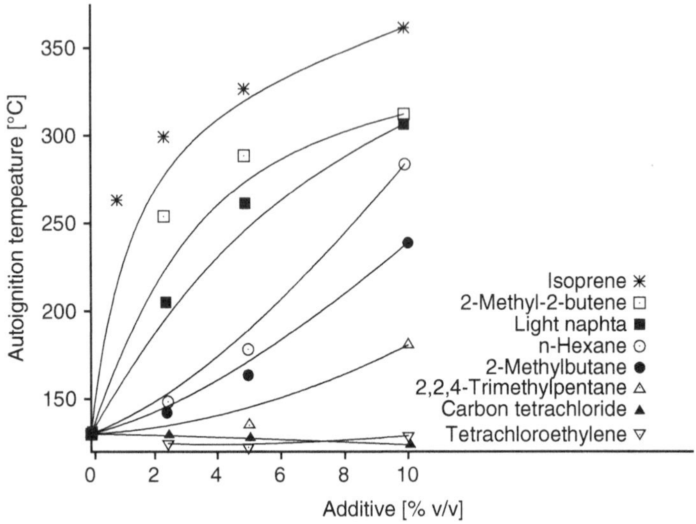

Figure 7.2 Additives to change the autoignition temperature of carbon disulfide. *(Source: Hyne JB, Greidanus JW. Additives for enhancing the autogenous ignition temperature of carbon disulphide. Can J Chem Eng 1970;48(4):47–13. http://dx.doi.org/10.1002/cjce.5450480423).*

Carbon disulfide has proven very effective because of its unique ability to solubilize even the most difficult of paraffins. For example, carbon disulfide is able to solubilize paraffins containing 20–30 carbon atoms. A drawback of carbon disulfide is the fact that it has a flash point of −30°C and an autoignition temperature of only 125°C. This drawback can be overcome when the carbon disulfide is generated in situ. Also, some additives can enhance the autoignition temperature of carbon disulfide [21]. This phenomenon is shown in Fig. 7.2.

Thiocarbonate salts, dithiocarbamates, and xanthic acids are compounds that are capable of generating carbon disulfide. An example is sodium dimethyldithiocarbamate [20]. This is unstable in an acidic environment.

7.3.2 Polyacrylamide

It has been shown that for oils which differ in the content of *n*-paraffins and asphalt-resinous substances, it is necessary to use blends of copolymers of different compositions and molecular weights to obtain optimal efficiency.

Polyacrylamide and wastes from the production of glycerol with a concentration of 400 mg per liter of oil have also been claimed to be effective as paraffin inhibitors [22].

7.3.3 Surfactants for Pigging

Usually, pigging operations are performed to remove the organic and inorganic debris from the walls of a pipeline. However, such scales may become very compact and may strongly adhere to the pipeline walls, so it is often necessary to add surfactants. The chemicals summarized in Table 7.2 have been tested.

The efficacy was assessed by solvency tests [18]. Examples of anionic surfactants for pigging are given in Table 7.3.

Table 7.2 Cleaning chemicals

Compound
Xylene
Methanol
Toluene
Specialty formulation: water soluble/oil miscible
Specialty formulation: water soluble/oil miscible with chelating agent
Specialty formulation: oil soluble/water miscible with odor control
Specialty formulation: oil soluble/water miscible with odor control and chelating agent
Glacial acetic acid 1%
Glacial acetic acid 25%

Source: Wylde JJ, Slayer JJL, Development, testing, and field application of a heavy-oil pipeline-cleaning chemical: a cradle-to-grave case history. SPE Proj Facil Construct 2010; 5(01). http://dx.doi.org/10.2118/119688-PA.

Table 7.3 Surfactants for pigging

Anionic surfactants	
Sodium lauryl sulfate	Sodium lauryl ether sulfate
Ammonium lauryl sulfosuccinate	Ammonium lauryl sulfate
Ammonium lauryl ether sulfate	Sodium dodecylbenzenesulfonate
Triethanolamine dodecylbenzenesulfonate	Sodium cocoyl isothionate
Sodium lauroyl isothionate	Sodium *N*-lauryl sarcosinate

Source: Smith KW, Schwartz KM. Gel plugs and pigs for pipeline use. 2007. http://www.freepatentsonline.com/7183239.html.

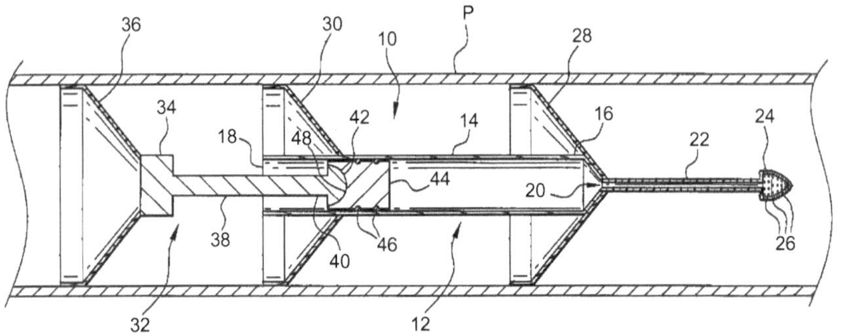

Figure 7.3 Apparatus for pigging. *(Source: Qanaei AAAJ. Chemical pigging apparatus for pipelines. US Patent 8719989; 2014. http://www.freepatentsonline.com/8719989.html).*

An exploded perspective view of a chemical pigging apparatus is shown in Fig. 7.3.

The chemical pigging apparatus 10 negotiates two sections of the pipeline. The forward component 12 of the chemical pigging apparatus includes a hollow cylindrical chemical chamber 14 having a first or forward end 16 and a second or rearward end 18. The forward end 16 is in communication with a relatively narrow diameter chemical dispensing passage, while the rearward end 18 is open. An elongated, relatively narrow diameter chemical dispensing tube 22 extends forward from the chemical dispensing passage 20 in communication with the forward end 16 of the chemical chamber 14, with the forward end of the dispensing tube 22 having a perforated chemical dispensing tip 24 extending therefrom to dispense the chemicals moved from the chemical chamber 14 through the chemical dispensing passage 20 and the chemical dispensing tube 22. The dispensing tip 24 has a diameter D2 that is relatively larger than the diameter D1 of the dispensing tube 22. This provides the dispensing tube 22 with relatively greater flexibility for negotiating bends and curves in a pipeline P, while providing a relatively sufficient surface area for a reasonable number of perforations or orifices 26 in the dispensing tip 24 [23].

7.3.4 Strong Acid

The use of a strong acid is the simplest way to dissolve such a deposit. But use of a strong acid generates large volumes of highly toxic hydrogen sulfide gas, which is an undesirable by-product. Addition of an oxidizing agent may avoid such toxicity hazards but oxidation products are then produced, including elemental sulfur, which is corrosive to pipes.

Another agent for treating such deposits is acrolein, but it also has health, safety, and environmental problems. Tris(hydroxymethyl)phosphine can solubilize iron and iron sulfide by forming a bright red water-soluble complex.

7.3.5 Chelating Agents

A blend of chelating agents that show a synergistic action has been described. These chelating agents are summarized in Table 7.4 and some are shown in Fig. 7.4.

Both iminodisuccinic acid and tetrakis(hydroxymethyl)phosphonium sulfate are environmentally friendly chelating agents [19].

It has been observed that the overall performance of the iron-complexing agents improves at alkaline pH. The pH of such a mixture can be adjusted to the alkaline range by use of ammonium hydroxide, ammonium chloride, ammonium citrate, ammonium lactate, ammonium acetate, potassium citrate, potassium hydroxide, potassium formate, sodium hydroxide, sodium acetate, or sodium formate.

Also, metal ions other than iron can form soluble complexes and be removed in the cleaning process. The presence of a surfactant helps in the dispersion and prevents deposits from reforming at downstream points within a pipeline network [19].

7.3.6 Enzymes

Enzymes from *Thauera aromatica* are active in the metabolism of toluene and all cresols, but not with any xylene isomers. Most of the aromatic

Table 7.4 Synergistic chelating agents

Compound
Polyaspartates
Hydroxyaminocarboxylic acid
Hydroxyethyliminodiacetic acid
Iminodisuccinic acid
Ethylenediaminetetracetic acid
Diethylenetriaminepentaacetic acid
Nitrilotriacetic acid
Tetrakis(hydroxymethyl)phosphonium sulfate

Source: Trahan DO. Method and composition to remove iron and iron sulfide compounds from pipeline networks. US Patent 8 673 834; 2014. http://www.freepatentsonline.com/8673834.html.

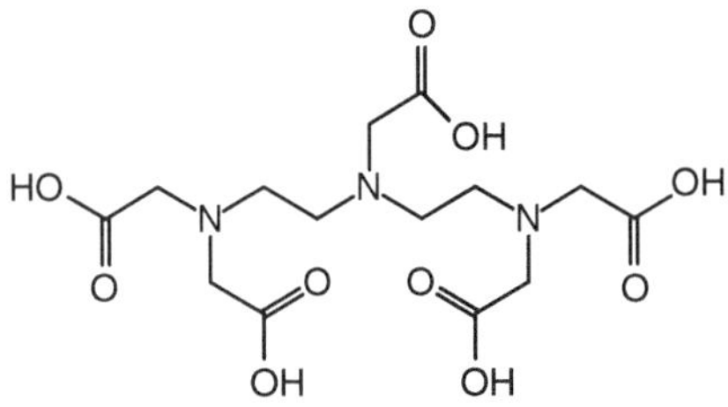

Diethylenetriaminepentaacetic acid

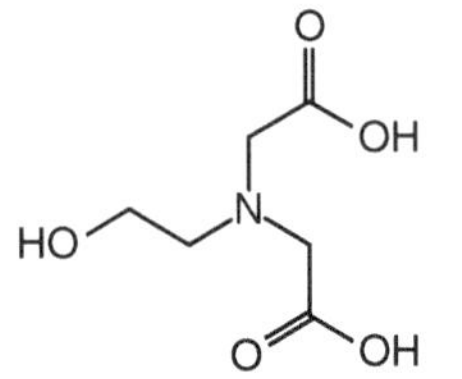

Hydroxyethyliminodiacetic acid

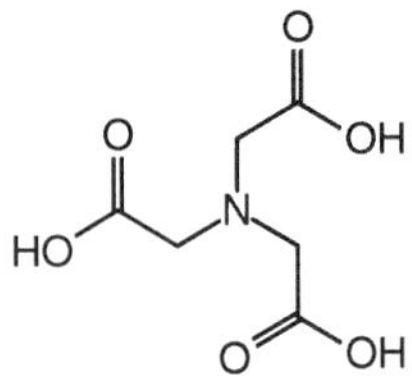

Nitrilotriacetic acid

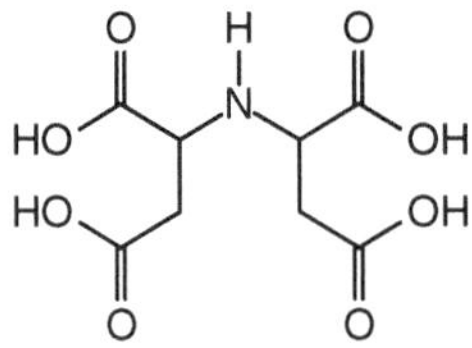

Iminodisuccinic acid

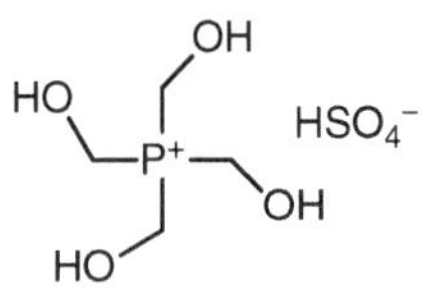

Tetrakis(hydroxymethyl)phosphonium sulfate

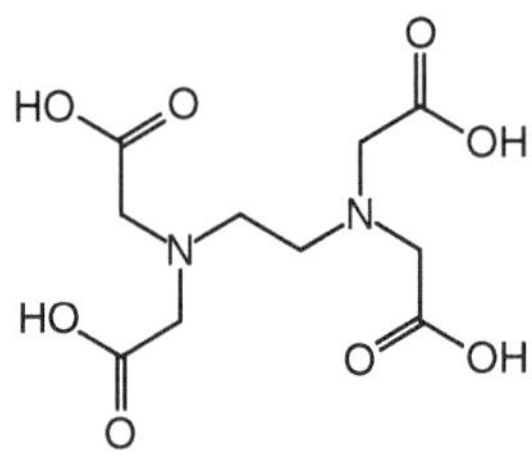

Ethylene diaminetetracetic acid

Figure 7.4 Chelating agents.

compounds are converted to the central intermediate benzoyl-CoA via different metabolic pathways. In enhanced oil recovery this strain may be useful in the maintenance of oil pipelines [24].

7.3.7 Gelled Pigs

Gelled fluid pigs will perform most of the functions of conventional pigs, but they have additional chemical capabilities. In addition, they can be injected into a pipeline through a valve. However, for displacement by a gas, gel pigs must be propelled by a mechanical pig.

Most pipeline gels are water based, but a variety of chemicals, solvents, and acids can be gelled. Gelled diesel, an organic gel, was first patented for pipeline use in 1973 [25]. The gels can be used for waste material removal, separation of products, placement of biocides and inhibitors, and removal of trapped mechanical pigs [26, 27].

An ablating gelatin pig has been described for use in pipelines. Because of the properties of gelatin, the pig will ablate, thereby depositing a protective layer onto the wall of the pipe [28]. The pig can be molded outside the pipe or it can be formed in situ.

The pig is formed by the mixing of gelatin with a heated liquid and then cooling of the mixture to ambient temperature. Preferably, the liquid contains a corrosion inhibitor or a drag reducer. In some applications, a slug of the treating solution is also passed through the pipeline between two ablating gelatin pigs. For high-temperature applications, a hardener may be added to increase the melting temperature of the pig.

Coacervate gels for pipeline pigs can be made from anionic or cationic polymers, a smaller amount of a surfactant having a charge opposite that of the polymer, and a hydrophobic alcohol [29]. Optional gel-promoting additives include betaines and amine oxides. An example of such a gel contains poly(diallyldimethylammonium chloride), sodium lauryl sulfonate, and lauryl alcohol.

Clarified diutan can be used as a gelling agent [30]. Diutan is a polysaccharide, which may be prepared by fermentation reactions. Diutan has glucose and rhamnose repeat units and a dirhamnose side chain [31, 32]. The preparation of a diutan gum with a strain of the microorganism *Xanthomonas campestris* has been described [33]. More recently, an improved synthesis of diutan has been described that involves the generation of a derivative of *Sphingomonas* sp. ATCC 53159 [34]. Such a modified diutan polysaccharide has proven particularly useful as a viscosifier in oilfield applications and within cement materials.

Clarified xanthan may also be used as the gelling agent [35]. Cross-linking may be advantageous when one is such fluids are used as a pig for pipeline cleaning. Suitable cross-linkers include boron derivatives, potassium periodate, and various metal complexes [30].

7.4 SCALE INHIBITION

Water from natural sources often contains dissolved minerals, with an appreciable presence of ions such as Ca^{2+}, Mg^{2+}, and, in the case of oilfield formation water, Ba^{2+}, Sr^{2+}, and also Ra^{2+} [36]. Under conditions of temperature or pH change, loss of carbon dioxide from solution or admixture with other water containing a different mineral content, relatively insoluble species such as carbonates and sulfates may deposit from solution as scale. In offshore oilfields such deposition may be particularly acute when sulfate-containing seawater, pumped underground to aid oil recovery, comes into contact with formation water.

The deposited scale may impede the oil recovery and may even become severe enough to block an oil well. Therefore a common procedure is to treat oil wells with a scale inhibitor to minimize or prevent scale deposition in the pipework.

Conventionally, a comparatively concentrated solution of the inhibitor is pumped down the oil well and flushed out into the formation. From here it leaches back into the water produced, thus protecting the well and the pipework from scaling.

The scale inhibitor should not only control the scale but must also be sufficiently soluble in the water at the temperatures it will meet to enable placement in the formation without itself prematurely precipitating from solution.

The retention of a scale inhibitor can be achieved by adsorption and precipitation processes. Precipitation is a known method for achieving longer scale inhibitor squeeze lifetimes.

A method for injecting an acidic solution containing a mixture of a scale inhibitor and multivalent and monovalent cations has been described [37]. Nitrilotri(methylene phosphonic acid) and diethylenetriaminepenta(methylene phosphonic acid) are scale inhibitors for such use. A multivalent cationic salt of the scale inhibitor is formed within the formation by ion exchange due to the buffering effect of the formation. This method is limited by the reservoir type and mineralogy and is often ineffective because of the poor yield of the precipitation reaction [36].

Another method for injecting an acidic scale inhibitor composition is to use a urea-type compound [38]. On placement downhole, the urea-type compound is thermally decomposed to form a basic solution. At the higher pH the cationic salt of the scale inhibitor is formed and precipitates. However, at lower temperatures of 40–60°C the urea compound will not decompose, so the pH will remain the same and the scale inhibitor salt will not precipitate. Therefore such a process is very temperature dependent and is limited to reservoirs with a bottom hole temperature greater than 80°C [36].

The injection of a scale inhibitor using a metal-chelating agent with a stability constant lower than that of the phosphonate scale inhibitor is also possible, together with a dissolved multivalent salt [39]. This results in the release of the multivalent cations from the chelating agent. The phosphonate inhibitor then precipitates as the cationic salt. The precipitation reaction occurs immediately on mixing of the inhibitor with the other components. A disadvantage is that the premature precipitation can cause a poor product placement in the reservoir. A delay in the precipitation reaction can be effected by alteration of the pH. The chelating agent must therefore have a lower stability constant than the phosphonate. At high well temperatures the scale inhibitor will precipitate too quickly and product placement will not be very accurate [36].

An improved composition contains a metal-chelating agent, a scale inhibitor, and divalent metal cations [36]. The stability constant of the metal-chelating agent–metal cation chelate at ambient temperature should be equal to or higher than the stability constant of the chelate formed. Further, the solubility of the chelate should decrease with increasing temperature.

Preferred chelating agents are shown in Tables 7.5 and 7.6. Scale inhibitors are shown in Table 7.7.

Table 7.5 Chelating agents

Compound	
α-Alanine	*N*'-tetraacetic acid
1,2-Diaminocyclohexane	2-Sulfoanilinediacetic acid
3-Sulfoanilinediacetic acid	4-Sulfoanilinediacetic acid
8-Hydroxyquinoline-5-sulfonic acid	Adenosine diphosphate
Adenosine phosphate	Adenosine triphosphate
Anilinediacetic acid	Aspartic acid
Gluconic acid	Glutamic acid
Glyceric acid	Glycine
Glycolic acid	Iminodiacetic acid
Lactic acid	Malic acid
Nitrilopropionic diacetic acid	Nitrilotriacetic acid
Nitroacetic acid	Salicylic acid
Serine	Tartaric acid
Tetrametaphosphoric acid	Trimetaphosphoric acid
Trimethylenediaminetetraacetic acid	Tripolyphosphoric acid
Tyrosine	Citric acid

Source: Heath S, Todd M. Scale inhibitor. US Patent 8 822 390, assigned to Clariant Finance (BVI) Limited (Tortola, VG); 2014. http://www.freepatentsonline.com/8822390.html.

Table 7.6 Chelating agents

Compound
N-Acetamidoiminodiacetic acid
2-Aminobenzoic acid-*N*,*N*-diacetic acid
3-Aminobenzoic acid-*N*,*N*-diacetic acid
4-Aminobenzoic acid-*N*,*N*-diacetic acid
Aminobarbituric acid-*N*,*N*-diacetic acid
β-Aminoethylphosphonic acid-*N*,*N*-diacetic acid
β-Aminoethylsulfonic acid-*N*,*N*'-tetracetic acid
Aminomethylphosphonic acid-*N*,*N*-diacetic acid
N-Benzylethylenediaminetriacetic acid
1,2-Bis(2(dicarboxymethyl)aminoethoxy)ethane
N-Butylethylenediaminetriacetic acid
N-Carbethoxy-β-aminoethyliminodiacetic acid
N-Cyanomethyliminodiacetic acid
N-Cyclohexylethylenediaminetriacetic acid
1,3-Diaminocyclohexane-*N*,*N*'-tetraacetic acid
Diethylenetriaminepentaacetic acid
Disodium 1,2-dihydroxybenzene-3,5-sulfonate
Ethylendiaminetetra(methylene phosphonic acid)
Ethylenediaminetetraacetic acid
Ethylenediamine-*N*,*N*-diacetic acid
N-Hydroxyethylenediaminetriacetic acid
N-Hydroxyethyliminodiacetic acid
β-Mercaptoethyliminodiacetic acid
N-Methoxyethyliminodiacetic acid
N-Methyliminodiacetic acid
N-Methylthioethyliminodiacetic acid
Pentamethylenediaminetetraacedic acid
Tetramethylenediaminetetraacetic acid
β-*N*-Trimethylammonium ethyliminodiacetic acid
N,*N*'-Ethylenebis(2(*o*-hydroxyphenyl))glycine
N,*N*-Dihydroxyethylethylenediaminediacetic acid

Source: Heath S, Todd M. Scale inhibitor. US Patent 8 822 390, assigned to Clariant Finance (BVI) Limited (Tortola, VG); 2014. http://www.freepatentsonline.com/8822390.html.

The performance of an exemplary composition is described as follows [36]: For a composition containing 3% $CaCl_2 \cdot 2H_2O$, 6% diethylenetriaminepentaacetic acid and 30% diethylenetriaminepenta(methylene phosphonic acid), the pH was adjusted with NaOH to between 4.4 and 6.2. The composition was placed in synthetic seawater. The stability constant for both the diethylenetriaminepentaacetic acid and the

Table 7.7 Scale inhibitors

Compound
Diethylenetriaminepenta(methylene phosphonic acid)
Nitrilo(methylene phosphonic acid) methacrylic diphosphonate homopolymer
Acrylic acid–allyl ethanolamine diphosphonate copolymer
Sodium vinyl sulfate–acrylic acid–allyl ammonia diphosphonate terpolymer
Acrylic acid–maleic acid–diethylenetriamine allyl phosphonate terpolymer
Polyaspartic acid
Polycarboxylates

Source: Heath S, Todd M. Scale inhibitor. US Patent 8 822 390, assigned to Clariant Finance (BVI) Limited (Tortola, VG); 2014. http://www.freepatentsonline.com/8822390.html.

diethylenetriaminepenta(methylene phosphonic acid) with calcium is 10^{10}. The solution was completely stable at room temperature for 60 days.

7.5 DEWATERING

After a pipeline for the transportation of light hydrocarbons such as natural gas has been repaired or constructed and hydro-tested or otherwise exposed to water, it is mandatory that water remaining in the pipeline be removed [40]. The reason for this is that light hydrocarbon gases form hydrates with water and these can and often do reduce or block the flow of gases through pipelines.

7.5.1 Alcohols

Conventionally, fluids such as methanol and glycols also in gelled form are used for the dewatering of pipelines [41]. For these chemicals the acceptable limitations for both subsea and overboard discharge may be exceeded.

Further, the methanol is usually separated in stages by pipeline pigs, which allows the water in the pipeline to be exposed to successive methanol contact, which brings about successive dilutions of the methanol-water mixture remaining in the pipeline [40]. Such methanol treatments may result in films of methanol-water mixtures remaining in pipelines containing very small amounts of water. However, even very small amounts of water may be sufficient to cause hydrate formation in very long, cold pipelines.

For example, a 24-in pipeline which is about 1100 km long will be left with a film of methanol and water mixture on its inside surfaces after a methanol pigging treatment of the type described above. Assuming the film is about 0.1 mm thick, the total volume of the methanol-water mixture left in the pipeline is about 220 m^3, most of which is methanol.

A disturbing property of methanol-water mixtures is that methanol evaporates faster than water and thus the water concentration of the mixture constantly increases.

7.5.2 Potassium Formate

An improved system for the dewatering and conditioning of pipelines is made from a concentrated potassium formate to provide hydrate control. To this composition a gelling agent (ie, a hydratable polymer) is added. Hydratable polymer are summarized in Table 7.8.

Guar gum is also called guaran. It is extracted from the seed of the leguminous shrub *Cyamopsis tetragonoloba*, where it acts as a food and water store. Structurally, guar gum is a galactomannan with a 1,4-linked β-D-mannopyranose backbone and branching points of galactose from the 6-positions. A higher galactose substitution increases the stiffness, but reduces the overall extensibility and radius of gyration of the isolated chains. The galactose residues prevent strong chain interactions.

A guar gum can be derivatized by reaction with propylene oxide or chloracetic acid, thus producing hydroxypropylguar and carboxymethyl-hydroxypropylguar, respectively. These reaction products exhibit enhanced hydration properties. The carboxyl functionality allows cross-linking at pH less than 7.

Guar gum is an economical thickener and stabilizer. It hydrates rapidly in cold water to give highly viscous pseudoplastic solutions of a greater low-shear viscosity in comparison with other hydrocolloids. Guar gum is more

Table 7.8 Hydratable polymers

Compound	
Hydropropyl guar	Hydroxypropyl cellulose
Carboxymethyl guar	Carboxymethylhydropropyl guar
Hydroxyethyl cellulose	Hydroxypropyl cellulose
Mannose and galactose polymers	Xanthan
Carboxymethyl hydroxyethyl cellulose	Scleroglucan
Polyacrylamide	Polyacrylate polymers

Source: Curr J, Hallett B, Sweeney A. Methods using formate gels to condition a pipeline or portion thereof. US Patent 8 746 044, assigned to Clearwater International LLC (Houston, TX); 2014. http://www.freepatentsonline.com/8746044.html.

soluble than locust bean gum and a better emulsifier as it contains more galactose branch points. Unlike locust bean gum, it does not form gels but does show good stability to freeze-thaw cycles.

The use of gelled metal formate compositions for the dewatering of pipelines proved to have an added benefit compared with the use of fluids such as methanol and glycols because of the formation of a gel column. In addition, the gel column established is compatible with all metal alloys and elastomers. Furthermore, one can reuse the gelled formate compositions by breaking the gel column, filtering the debris out of the resulting fluid, and regelling the recovered formate solution with or without the adjustment of formate concentration or pH.

7.5.3 Azeotropic Mixtures

An improved method for drying pipelines containing water is to allow a flow of a liquid that forms an azeotropic mixture with the water in the pipeline. When this mixture is evaporated in the pipeline, the water concentration in the remaining mixture does not approach a concentration which could form hydrates. Examples of azeotropic mixture-forming liquids are summarized in Table 7.9.

Generally, azeotropic liquid alcohols, such as ethanol, are the preferred compounds. The azeotropic mixtures formed with water in the pipeline readily evaporate at a constant ratio of azeotropic liquid to water. Further, the presence of an azeotropic liquid alcohol in the remaining unevaporated mixture depresses the temperature at which hydrates will form with the water in the mixture [40]. Thus, if we ensure that the amount of water in the azeotropic mixture initially remaining in the pipeline is low, the evaporation of the mixture will not result in a high enough ratio of water to form hydrates with the hydrate-forming gaseous fluid flowing through the pipeline.

Table 7.9 Liquids that form azeotropes with water

Compound	
Ethanol	1-Propanol
2-Propanol	*tert*-Butyl alcohol
Isopropyl ether	Ethyl acetate
Methyl ethyl ketone	

Source: Bradley GW. Methods of drying pipelines. US Patent 6 076 278, assigned to Halliburton Energy Services, Inc. (Duncan, OK); 2000. http://www.freepatentsonline.com/6076278.html.

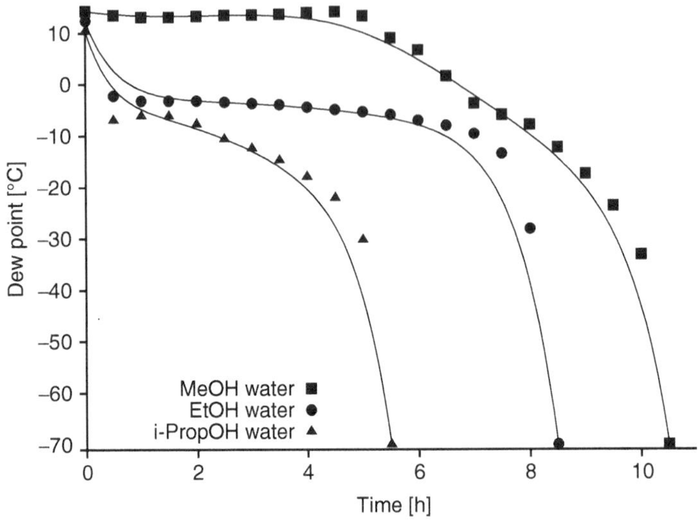

Figure 7.5 Dew points with time. *(Source: Bradley GW. Methods of drying pipelines. US Patent 6 076 278, assigned to Halliburton Energy Services, Inc. (Duncan, OK); 2000. http://www.freepatentsonline.com/6076278.html).*

A preferred technique for flowing the azeotropic liquid is to flow the liquid through the pipeline in stages separated by one or more pipeline pigs. The pigs are propelled through the pipeline by a high-pressure stream of non–hydrate-forming gas such as air, nitrogen, or an available process gas. The stream of non–hydrate-forming gas should have been previously dried.

Tests were performed in which quantities of an azeotropic mixture of alcohol and water were placed in a test apparatus. Dry nitrogen was passed through the apparatus and the dew points of the exiting nitrogen were recorded. The development of the dew points with time for various alcohols is shown in Fig. 7.5.

It can be seen from Fig. 7.5 that 2-propanol (isopropyl alcohol) has the best performance.

7.6 DEBRIS REMOVAL IN SUBSEA PIPELINES

The transportation process through a pipeline may be hindered as a result of deposition of debris. The presence of debris within the pipeline leads to a reduction in the effective flow area, thus reducing the efficiency of the transportation system [42].

Debris may be formed by changes in temperature or pressure of the transported fluid and in addition by chemical reactions. Pigging is performed routinely to extend the operational life of pipelines. A combination of mechanical and chemical remedial processes showed beneficial results in pipeline cleaning. Mechanical brush pigs loosen the debris adhering to the walls, while chemical techniques, such as the use of gelled fluids, enhance the carrying capacity and are helpful in the transport of the loosened debris out of the pipeline.

A fast-breaking gelled fluid for debris removal has been developed (ie, a highly viscous, aqueous-based linear polysaccharide gel). Both laboratory tests and field tests were performed to assess the properties of the gels investigated. The field test results indicated that the combined mechanical and gel system removed a significant amount of debris [42].

REFERENCES

[1] Quarini J, Shire S. A review of fluid-driven pipeline pigs and their applications. Proc Inst Mech Eng E 2007;221(1):1–10. http://dx.doi.org/10.1243/0954408JPME108.

[2] Heinze LR, Shahreyar N, Baruah BM. A review of past 50 years of paraffin prevention and removal techniques as presented in the SWPSC [Southwestern Petroleum Short Course]. In: Proceedings Volume, 48th Annu. Southwestern Petrol. Short Course Ass. Inc et al Mtg. (Lubbock, TX, 4/25–26/2001); 2001. p. 230–8.

[3] Duncum SN, James K, Osborne CG. Wax deposit inhibitors. US Patent 6 140 276, assigned to BP Exploration Operating Company Limited (London, GB); 2000. http://www.freepatentsonline.com/6140276.html.

[4] Sharov AG, Terteryan RA, Burova LI, Shulgaitser LA, Effect of a copolymer inhibitor of paraffin deposits on oils of various fields. Neft Khoz 1989;9:55–8.

[5] Sanjay M, Simanta B, Kulwant S. Paraffin problems in crude oil production and transportation: a review. SPE Prod Facil 1995;10(01):50–4. http://dx.doi.org/10.2118/28181-PA.

[6] Coto B, Martos C, Espada JJ, Robustillo MD, Peña JL, Experimental study of the effect of inhibitors in wax precipitation by different techniques. Energy Sci Eng 2014;2(4):196–203. http://dx.doi.org/10.1002/ese3.42.

[7] Sheikh Mohammad Samiur R, Chacko S, Improved paraffin-deposition-profile estimation in hydrocarbon pipelines and effective mitigation review. Oil Gas Facil 2013;2(1):78–85. http://dx.doi.org/10.2118/165933-PA.

[8] Holder GA, Winkler J. Wax crystallization from distillate fuels. 2: Mechanism of pour depression. J Inst Petrol 1965;51(499):235.

[9] Garcia MdC, Carbognani L, Asphaltene–paraffin structural interactions. effect on crude oil stability. Energy Fuels 2001;15(5):1021–27. http://dx.doi.org/10.1021/ef0100303.

[10] Akbarzadeh K, Ratulowski J, Eskin D, Davies T. The importance of wax-deposition measurements in the simulation and design of subsea pipelines. SPE Proj Facil Construct 2010;5(2):49–57. http://dx.doi.org/10.2118/115131-PA.

[11] Schlumber Software. Olga dynamic multiphase flow simulator. Internet; 2014. http://www.software.slb.com/products/foundation/Pages/olga.aspx.

[12] Hunt EB. Laboratory study of paraffin deposition. J Petrol Tech 1962;14(11):1259–69. http://dx.doi.org/10.2118/279-PA.
[13] Jorda RM. Paraffin deposition and prevention in oil wells. J Petrol Tech 1966;18(12):1605–12. http://dx.doi.org/10.2118/1598-PA.
[14] Mahto V, Verma D, Kumar A, Sharma VP. Wax deposition in flow lines under dynamic conditions. Petrol Sci Tech 2014;32(16):1996–2003. http://dx.doi.org/10.1080/10916466.2012.733468.
[15] Lazar I, Voicu A, Nicolescu C, Mucenica D, Dobrota S, Petrisor IG, et al. The use of naturally occurring selectively isolated bacteria for inhibiting paraffin deposition. J Petrol Sci Eng 1999;22(1–3):161–9. http://www.sciencedirect.com/science/article/pii/S0920410598000655, http://dx.doi.org/10.1016/S0920-4105(98)00065-5.
[16] Hameed MS, Taha EJ, Al-Jarrah MMF. Observations on the effect of low molecular weights polyethylene-oxides on the flow of water in closed pipes and standard fittings. J Petrol Res 1989;8(1):47–60.
[17] Khan HU, Handoo J, Agrawal KM, Joshi GC. Determination of wax separation temperature of crude oils from their viscosity behaviour. Erdöl Erdgas Kohle 1991;107(1):21–2.
[18] Wylde JJ, Slayer JJL, Development, testing, and field application of a heavy-oil pipeline-cleaning chemical: a cradle-to-grave case history. SPE Proj Facil Construct 2010;5(01). http://dx.doi.org/10.2118/119688-PA.
[19] Trahan DO. Method and composition to remove iron and iron sulfide compounds from pipeline networks. US Patent 8 673 834; 2014. http://www.freepatentsonline.com/8673834.html.
[20] McElligott Jr PJ, Sandler SR, Kennoy JF, Tuszynski WJ. Method for dewaxing oil and gas pipelines or wells. US Patent 5 053 149, assigned to Atochem North America, Inc. (Philadelphia, PA); 1991. http://www.freepatentsonline.com/5053149.html.
[21] Hyne JB, Greidanus JW, Additives for enhancing the autogenous ignition temperature of carbon disulphide. Can J Chem Eng 1970;48(4):471–3. http://dx.doi.org/10.1002/cjce.5450480423.
[22] Shmakova-Lindeman OE. Paraffin inhibitors. US Patent 7 417 009, assigned to Nalco Company (Naperville, IL); 2008. http://www.freepatentsonline.com/7417009.html.
[23] Qanaei AAAJ. Chemical pigging apparatus for pipelines. US Patent 8 719 989; 2014. http://www.freepatentsonline.com/8719989.html.
[24] Hendrickson ER, Jackson RE, Keeler SJ, Luckring AK, Perry MP, Wolstenholme S. Identification, characterization, and application of thauera sp. al9:8 useful in microbially enhanced oil recovery. US Patent 7 708 065, assigned to E.I. du Pont de Nemours and Company (Wilmington, DE); 2010. http://www.freepatentsonline.com/7708065.html.
[25] Purinton Jr RJ, Mitchell S, Practical applications for gelled fluid pigging. Pipe Line Gas Ind 1987;66(3):55–6.
[26] Kennard MA, McNulty JG, Conventional pipeline-pigging technology: Pt.2: Corrosion-inhibitor deposition using pigs. Pipes Pipelines Int 1992;37(4):14–20.
[27] Messner SF. Cleaning of pipelines with gel pigs (csotavvezetek tisztitasa geles csomalacokkal). Koolaj Foldgaz 1991;24(7):219–22.
[28] Lowther FE. Method for treating tubulars with a gelatin pig. US Patent 5 215 781, assigned to Atlantic Richfield Company (Los Angeles, CA); 1993. http://www.freepatentsonline.com/5215781.html.
[29] Smith KW, Schwartz KM. Gel plugs and pigs for pipeline use; 2007. http://www.freepatentsonline.com/7183239.html.
[30] Welton TD, Pauls RW, Song L, Bryant JE, Beach SR, Robb ID. Treatment fluids comprising diutan and associated methods. US Patent 7 960 315, assigned to Halliburton Energy Services, Inc. (Duncan, OK); 2011. http://www.freepatentsonline.com/7960315.html.

[31] Diltz S, Zeller SG, Location of o-acetyl groups in s-657 using the reductive-cleavage method. Carbohyd Res 2001;331(3):265–70. http://www.sciencedirect.com/science/article/pii/S0008621501000453, http://dx.doi.org/10.1016/S0008-6215(01)00045-3.

[32] Eoff LS, Pauls RW. Sealant compositions comprising diutan and associated methods. US Patent 8 343 896, assigned to Halliburton Energy Services, Inc. (Duncan, OK); 2013. http://www.freepatentsonline.com/8343896.html.

[33] Peik JA, Steenbergen SM, Veeder GT. Heteropolysaccharide s-657. US Patent 5 175 278, assigned to Merck & Co., Inc. (Rahway, NJ); 1992. http://www.freepatentsonline.com/5175278.html.

[34] Harding NE, Patel YN, Coleman R, Matzke S. High viscosity diutan gums. US Patent 7 868 167, assigned to CP Kelco U.S., Inc. (Atlanta, GA); 2011. http://www.freepatentsonline.com/7868167.html.

[35] Welton TD, Sanders MW, Mundy JL, Beach SR, Pauls RW, Harris PC, et al. Treatment fluids comprising clarified xanthan and associated methods. US Patent 7 825 073, assigned to Halliburton Energy Services, Inc. (Duncan, OK); 2010. http://www.freepatentsonline.com/7825073.html.

[36] Heath S, Todd M. Scale inhibitor. US Patent 8 822 390, assigned to Clariant Finance (BVI) Limited (Tortola, VG); 2014. http://www.freepatentsonline.com/8822390.html.

[37] Lawson JB, Richardson EA. Treating wells with ion-exchange-precipitated scale inhibitor. US Patent 4 393 938, assigned to Shell Oil Company (Houston, TX); 1983. http://www.freepatentsonline.com/4393938.html.

[38] Hen J. Inhibition of scale formation from oil well brines utilizing a slow release. US Patent 5 141 655, assigned to Mobil Oil Corporation (Fairfax, VA); 1992. http://www.freepatentsonline.com/5141655.html.

[39] Carlberg BL, Wolf NO, Pober KW, Nash WD. Inhibition of reservoir scale. US Patent 4 860 829, assigned to Conoco Inc. (Ponca City, OK); 1989. http://www.freepatentsonline.com/4860829.html.

[40] Bradley GW. Methods of drying pipelines. US Patent 6 076 278, assigned to Halliburton Energy Services, Inc. (Duncan, OK); 2000. http://www.freepatentsonline.com/6076278.html.

[41] Curr J, Hallett B, Sweeney A. Methods using formate gels to condition a pipeline or portion thereof. US Patent 8 746 044, assigned to Clearwater International LLC (Houston, TX); 2014. http://www.freepatentsonline.com/8746044.html.

[42] Misra C, Danait A, Salgaonkar L. Design of gelled fluid for debris removal in subsea pipeline. Oil Gas Facil 2013; (1):46–51. https://www.spe.org/ogf/print/subscribers/2013/02/16_Paper153157.pdf.

CHAPTER 8

Safety Aspects for Pipelines

8.1 ODORIZATION

The hazards of odorless combustible gases were possibly first realized by miners. The idea of odorization of combustible gases results from Julius Quaglio [1] in 1880, who investigated various aspects of water gas. The gas produced at that time contained impurities that caused the typical odor of gas. Later, the gas was obtained in such a purity that dangers emerged in handling. Some aspects concerning the history of this issue are given in the literature [2]. The primary objective of gas odorization is safety. Odorization serves as a warning in the detection of natural gas in air before it reaches combustible levels. Naturally, odorization is part of the risk management for natural gas pipelines [3].

Certain federal pipeline safety regulations, and also those of the US National Fire Protection Association, require that combustible gases in pipelines be detectable at one-fifth of the lower explosive limit by a person with a normal sense of smell, either by the natural odor of the gas or by means of artificial odorization [4]. Therefore the proper odorization and odorants are integral parts of safety [5, 6].

Odorization is a primary concern for any gas transmission company [5, 7]. Accurate injection of the odorant, proper monitoring techniques, and complete record maintenance are important factors in developing and sustaining a successful odorization program.

The aspects of odorization have been reviewed. Important points to consider are which pipelines require odorization, the detectable limits of gas odor, odorants and odorizing considerations, and monitoring a pipeline system to ensure that the odorization program is meeting the regulatory requirements [4].

Guide to the Practical Use of Chemicals in Refineries and Pipelines
http://dx.doi.org/10.1016/B978-0-12-805412-3.00008-8

8.1.1 General Aspects

Explosive Limits

The lower explosive limits of certain gases are shown in Table 8.1. The data in Table 8.1 are approximate data for ordinary air. There is also an upper explosive limit. The explosive limits are dependent on the oxygen content, which is, of course, constant for ordinary air; they are also dependent on the total pressure [8]. There are mathematical models to calculate the explosive limits [9].

For example, the program GasEq® and some additional software have been used to calculate the flammability limits of mixtures with, for example, ethylene oxide, air, and inert gases at temperatures between 20°C and 100°C and pressures between 0.4 and 1.0 bar.

GasEq® is a Microsoft Windows-based program with a Microsoft Excel interface. The program can be used to calculate the equilibrium of combustion and is intended primarily for gas-phase calculations, although there is a limited facility for condensed phases, such as soot [10].

Desirable Properties of Odorants

An ideal odorant should have the following desirable physical and functional properties [11]:

- low perceptual threshold,
- preferably distinguishable from smells in daily life and functionable as a warning smell,
- low boiling point (essential for hydrogen gas),
- low corrosivity,
- little or no olfactory fatigue,
- low toxicity.

Odorants should allow the detection of leakages without any external equipment. The end user needs not worry about maintaining any measuring equipment. Thus, equipment failures cannot result in leaks not being

Table 8.1 Lower explosive limits

Gas	Lower explosive limit (%)
Methane	4.4
Natural gas	3.9
Hydrogen	4
Liquid petroleum gas	1.9

detected. Consequently, odorants allow the detection of leaks in places where it may be difficult to position detectors. In addition, odorants can be used in low concentrations because of the sensitivity of the human olfactory system [12].

Furthermore, a fuel gas odorant should be easily distinguishable from smells in daily life for it to be suitable as a warning smell. Smells in daily life are ordinary smells that are experienced in daily life situations and are not perceived as a foreign or unusual odor. On the other hand, a warning smell is, in general, an unpleasant smell that is perceived as an odor indicating unusual situations, clearly distinguishable from the smells in daily life. In this way, the smell acts as a warning signal [11].

8.1.2 Measurement and Odor Monitoring

The odor monitoring methods have been reviewed by Klusman [13] and Wetteman and Wilson [14].

In the early days of the coal mining industry, open flames were taken into the mine. When the candle started burning irregularly, the possible danger of an explosion could be realized. However, this technique often indicated a danger too late. A significant step forward was the invention of the safety lamp in 1815, by Sir Humphry Davy. The presence of combustible gases could be still detected, but the explosion reaction was restricted to inside the wired cage of the safety lamp.

Another monitoring method utilized the extraordinary sensitivity of canaries to methane. Canaries accompanied the miners into the mines, admittedly involuntary in small cages. The canary stops singing in the presence of even small amounts of coal gas, and eventually dies at higher levels that are still harmless for the human body [12].

By the addition of odorants to liquid petroleum gas and natural gas an improved level of safety is achieved. However, the use of odorants suffers from certain limitations and disadvantages. Namely, for an odor to be detected, a human must be in the vicinity of the leak in order to detect the odor. A further drawback is that not every individual is able to detect the odors at the same mandatory level. Most dangerous is that some individuals are not able detect the odor at all. For example, the sensitivity of an individual may be affected by a seasonal illness such as influenza or a sniffle. Further, exposure to other odors lowers the sensitivity, and during sleep the olfactory response decreases.

Olfactory Response

Odorants are chemicals that stimulate the sense of smell. The human olfactory system is much less sensitive in comparison with the systems of other animals, such as dogs, which are notorious for their ability to smell. Nevertheless, humans can still detect certain odorants in concentrations in air in the parts per trillion range.

There are several definitions to quantify the odor. The threshold odor concentration is the absolute perception threshold at which a substance can be barely identified, but the impression is too faint. The odor recognition threshold is the concentration at which the sensation indicates a representative odor for a certain substance [15].

There is a difference between the minimum detectable concentration and the minimum identifiable concentration. The minimum concentration that is detectable is defined as the lowest concentration at which 50% of the human population is able *to smell something*. This concentration is sometimes called the perceptual threshold or odor recognition threshold.

In contrast, the recognition threshold is the minimum concentration at which a predefined percentage of the population can identify the substance coarsely. At the detection and recognition thresholds there will be a large number of individuals who do not detect or recognize the odor. The data indicate as a rule of the thumb that the recognition threshold is roughly a factor of 10 higher than the detection threshold [12].

Perceptual Threshold and Olfactory Intensity

The perceptual threshold of an odorant is the minimum concentration of the odorant in air, expressed, for example, in parts per million, at which a human can easily notice the smell of the odorant. In general, the values refer to ratios of volume and not to ratios of weight, as usual in gas analysis.

This threshold can be determined by a couple of panelists who assess the olfactory intensity of a test substance in an odorless chamber. The air in the chamber is stirred until the concentration of the test substance becomes constant and is then left to stand. The olfactory intensity is assessed, for example, on a scale of 0 to 5 for smell pollution. The perceptual and recognition thresholds of selected chemicals are summarized in Table 8.2.

The olfactory response of humans (males and females between the ages of 16 and 82 years) was tested with various odorants, including *tert*-butyl mercaptan, thiophene, ethyl mercaptan, dimethyl sulfide, isopropyl mercaptan, and mixtures of these odorants. The goal was to establish the

Table 8.2 Perceptual and recognition thresholds of chemicals

Perceptual thresholds [11]	**Concentration**
p-Cresol	51.3 ppt
5-Ethylidene-2-norbornene	4 ppb
Methyl mercaptan	1.6 ppb
γ-Undecalactone	22.8 ppt
3-Hydroxy-4-methyl-5-ethyl-2(5*H*)-furanone	7.4 ppt
trans-2-*trans*-4-Decadienal	87 ppt
Recognition thresholds [15, p. 14]	**Concentration**
Ethane	1500 ppm
Propane	11,000 ppm
Butane	5000 ppm
Pentane	900 ppm
Octane	200 ppm
Methanol	6000 ppm
Ethanol	6000 ppm
Octanol	2 ppb
Geosmin	5 ppt
Diethyl ether	300 ppb
Ethyl acrylate	2 ppb
Hydrogen sulfide	1 ppm
Methyl mercaptan	35 ppb
Ethyl mercaptan	2 ppb
Butyl mercaptan	0.8 ppb
Ethyl sulfide	4 ppb

Source: Verschueren K. Handbook of environmental data on organic chemicals. 5th ed. Hoboken, NJ: Wiley & Sons; 2009.

warning levels below the explosive limit in the event of a gas leak [16]. The study suggests that ethyl mercaptan is the most suitable odorant. Trained dogs can detect odorizing agents in concentrations as low as 10–18 ppb [17–19].

Odor Index

The odor index is the ratio of the vapor pressure to the odor recognition threshold [15, p. 13]:

$$\text{Odor index} = \frac{\text{vapor pressure (ppm)}}{\text{recognition threshold (ppm)}}. \tag{8.1}$$

Thus the odor index is a dimensionless number. An odor index less than 1 means that the substance has a vapor pressure that is too small for the substance to be smelled.

Olfactory Power

Another definition of the intensity of a certain smell is the olfactory power [20]. The olfactory power is defined as the negative decadic logarithm of the detection threshold. Olfactory power values are given in Table 8.3.

Table 8.3 Functionality and odor potency of low molecular weight compounds

Functionality	Compound	Olfactory power
Hydrocarbons	Ethane	2.00
	Propane	2.57
	Butane	3.69
Halides	Chloromethane	4.99
	Ethyl chloride	5.39
Alcohols	Methanol	3.85
	Ethanol	4.54
	1-Propanol	5.62
Esters	Methyl formate	4.03
	Methyl acetate	5.21
Ketone	Acetone	4.84
Aldehydes	Formaldehyde	6.06
	Acetaldehyde	6.73
Amines	Methylamine	7.73
	Dimethylamine	7.09
	Ethylamine	6.49
	Diethylamine	6.73
	Propylamine	7.96
Thiols	Methyl mercaptan	8.98
	Ethyl mercaptan	8.97
	Isobutyl mercaptan	8.95
	tert-Butyl mercaptan	9.48
Sulfides	Dimethyl sulfide	8.65
	Methylethyl sulfide	8.42
	Diethyl sulfide	8.41
Selenide	Diethyl selenide	9.13
Selenol	Ethylselenol	10.74

Source: Flynn PJ, Sprague MJ. Hydrogen odorants and odorant selection method. US Patent 7 811 688, assigned to Enersol, Inc. N.A., LP (Fairfax, VA); 2010. http://www.freepatentsonline.com/7811688.html.

Physiological Methods

An exemplary specific method for determining the perceptual threshold is as follows. A test odorant in a dish is left in an odorless chamber for a given period of time. After agitation of the air in the odorless chamber until the concentration of the test substance becomes constant followed by it standing for 1 min, panelists enter the chamber and assess the olfactory intensity on a scale of 0 to 5 [11]:

0: Odorless,
1: Slight smell, but not identified,
2: Easily noticed and can be identified,
3: Obvious smell,
4: Strong smell,
5: Intolerably strong smell.

This procedure is repeated at different concentrations of the test odorant. The perceptual threshold of the test odorant can be obtained by determination of the concentration of the odorant corresponding to an olfactory intensity of 2, at which the smell can easily be identified.

Triangle Odor Bag Method

The measurement of the odor threshold by the triangle odor bag method has been described in detail [21]. The data were collected over a comparatively long period (ie, from 1976 to 1988). Two hundred twenty-three substances were tested.

Isoamyl mercaptan exhibited the lowest threshold of 0.77 ppt and propane exhibited the highest threshold of 1500 ppm. The distribution of the thresholds of the substances investigated follows a Gaussian normal distribution. As expected it was established that sulfur compounds apart from sulfur dioxide and carbon disulfide exhibit a comparatively low threshold.

There is a relation between the molecular weight of the substance and its odor threshold. There is a great difference in the thresholds between isomers. The thresholds may differ by up to four powers of 10. An extensive list of the odor thresholds measured in the study described has been presented [21]. For comparison, in Table 8.4 selected odor thresholds measured by the triangle odor bag method are reproduced.

The values presented in Table 8.4 differ in some cases appreciably from those given in Table 8.2. This may be caused by the different methods used in the acquisition. Actually, in the study, repeated tests for the same substances were performed to some extent. Within these tests a considerable variation was reported.

Table 8.4 Odor thresholds measured by the triangle odor bag method

Substance	Concentration
Propane	1500 ppm
Butane	1200 ppm
Pentane	1.4 ppm
Octane	1.7 ppm
Methanol	33 ppm
Ethanol	0.52 ppm
Octanol	2.7 ppb
Methyl mercaptan	70 ppt
Ethyl mercaptan	8.7 ppt
n-Propyl mercaptan	13 ppt
Isopropyl mercaptan	6 ppt
n-Butyl mercaptan	2.8 ppt
Isobutyl mercaptan	6.8 ppt
Butyl mercaptan	30 ppt
tert-Butyl mercaptan	29 ppt
n-Amyl mercaptan	0.78 ppt
Isoamyl mercaptan	0.77 ppt
n-Hexyl mercaptan	15 ppt

Source: Nagata Y. Measurement of odor threshold by triangle odor bag method. In: International symposium on odor measurement, Asian Network on Odor Measurement and Control. Tokyo, JP: Japan Association on Odor Environment; 2003. p. 118-Ű27. [electronic:] http://www.env.go.jp/en/air/odor/measure/0232.pdf.

Standardized Methods

Sensory thresholds have been established in order to determine the potential of substances at low concentrations to impart odor, taste, haptic properties, etc., to some form of matter [22].

Procedures for referencing the odor intensities of odorous materials have been standardized. The so-called *ASTM Odor Intensity Referencing Scale* has been developed [23]. This scale is a geometric progression scale.

Reference odorant vapors are presented for evaluation by a panel of at least eight independent judges. The test relies on comparison of the odor intensity of the sample to the odor intensities of a series of concentrations of a reference odorant, which is 1-butanol.

Two methods that create the smell are in use. In the dynamic scale method, a dynamic-dilution apparatus is used, which is equipped with a series of sniffing ports from which constant concentrations of 1-butanol emerge at constant volumetric flow rates in air. In the static-scale method, a series of Erlenmeyer flasks containing known concentrations of 1-butanol dissolved in water are used.

Chemical and Physical Methods

Chromatographic and Spectroscopic Methods

The concentration of odorants in gases can be measured by the absorbance in the ultraviolet region [24]. The absorbance of odorized gas is much higher than the absorbance of untreated gas.

An integrated natural gas pipeline leak detector has been developed [25]. The detector is based on near infrared diode laser absorption spectroscopy, wavelength modulation spectroscopy, and harmonic detection. Direct absorption spectroscopy has the drawback of low sensitivity.

It is difficult to detect a small flux pipeline leak of natural gas. Therefore to improve the sensitivity, the second harmonic signal is used in a modulated wave. The method accesses directly the response of methane, so basically the presence of an odorant is immaterial.

Gas chromatography with an electrochemical detector is also suitable [26] for analysis of mercaptans. Portable gas chromatographic equipment for the measurement and analysis of odorants in gas distribution networks is available.

Besides applications in the field of petroleum, the analysis and characterization of odorants plays an important role in the food industry as well in the perfume industry. Remarkably, the sensitivity of an analytical instrument (eg, a mass spectrometer) and the sensitivity of the human nose do not correlate at all. Substances with an intense smell may cause only a small mass spectrometric peak, and vice versa.

Colorimetric Methods

A classical analysis of traces of gases consists in the use of colorimetric methods. These methods were originally invented by Dräger in Lübeck, Germany. A test tube consists of a thin glass tube that is closed at both ends by melting of the glass. Inside the tube, an indicator is placed that reacts with the substance to be analyzed, resulting in a change in the color.

At probing, the molten ends of the glass tube are broken and a known amount of gas is sucked through the test tube. Depending on the concentration of the gaseous impurity under consideration, a certain distance becomes colored. On the tube itself, a scale is printed so that the concentration of the respective impurity can be read out directly in parts per million, provided the correct volume of gas has been sucked before.

Electronic Nose

The term *electronic nose* was created around the late 1980s [27]. The principle of the electronic nose is based on a chemical sensor [28]. A chemical sensor is a device that converts chemical information into an analytically useful signal.

Chemical sensors are important for a variety of industrial and environmental applications, including the detection of hazardous chemicals, quality control in the food, perfume, and beverage industries, and medical applications [29]. These types of sensors include a sorbent layer deposited on an active area of a signal transducer.

Chemicals sensors have been described that are, among others, potentially suitable for the detection of natural gas [30, 31]. For example, conducting polypyrrole films have been found to be useful for the detection of methane. The sensitivity of a polypyrrole film is dependent on the methane concentration and the pressure.

Nanocomposites based on iron oxide and polypyrrole have been used for the sensing of various gases [32]. These composites can be used for the sensing of methane in oilfields, natural gas pipelines and joints, and wastewater treatment plants [30].

Commercial metal oxide-based gas sensors that are specifically sensitive to methane are available. They are recommended as sensors for domestic gas alarms for the detection of methane [33].

On the other hand, sensor systems have been described for the detection of water traces in natural gas [34]. They operate by the change of the electric capacitance caused by the presence of water vapor.

8.1.3 Additives for Odorization

Sulfur Compounds

Because of their inherent penetrating smell, certain organic sulfur compounds are used for odorization. Repellents from the skunk contain compounds such as *trans*-2-butene-1-thiol and 3-methyl-1-butanethiol. Ethyl mercaptan, because of its extremely low odor threshold, is the favorite compound used as an odorant in natural gas and liquid propane for leak detection. Tetrahydrothiophene is also often used. Common odorization reagents are summarized in Table 8.5 and Figs. 8.1 and 8.2. Selected properties of the most important compounds are given in Table 8.6.

Table 8.5 Sulfur-based additives for odorization

Additive	References
Ethyl mercaptan or a mixture of ethyl, propyl, and butyl mercaptans, dimethyl disulfide, diethyl disulfide, and methyl ethyl disulfide[a]	[35, 36]
Diethyl sulfide and ethyl propyl sulfide	[37]
Cyclohexyl mercaptan	[38]
Dimethyl sulfide	[39]
Mixture of ethyl, propyl, butyl, and amyl mercaptans	[40]
2-Methoxy-3-isobutylpyrazine and 4-methyl-4-mercapto-2-pentanone	[40]

[a]From wastes.

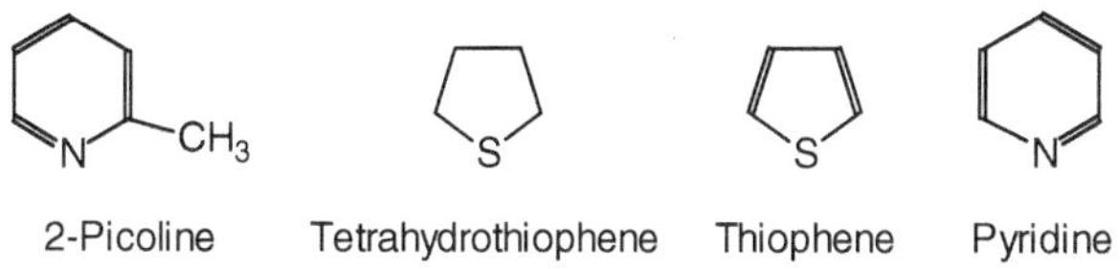

Figure 8.1 Cyclic odorants.

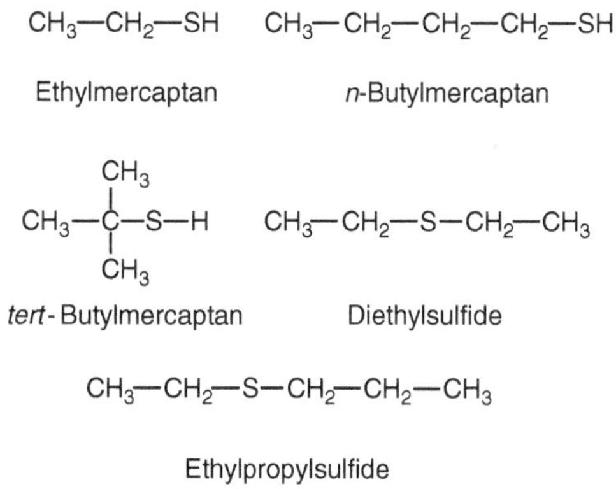

Figure 8.2 Mercaptans and sulfides.

tert-Butyl mercaptan is very common in a single-component odorization method. It has good soil penetration and high resistance to oxidation. A drawback is its high freezing point, which suggests it should be used in mixtures with other components.

Isopropyl mercaptan has a strong odor and also good resistance to oxidation. It is usually blended with *tert*-butyl mercaptan to depress the freezing point of the latter [2].

Table 8.6 Properties of odorization additives

Additive	Boiling point (°C)	Odor threshold (ppb)
Methyl mercaptan	4	0.11
Ethyl mercaptan	35	0.019
Propyl mercaptan	53	0.0075
tert-Butyl mercaptan	63	0.008

Source: O'Rear DJ. Use of waste nitrogen from air separation units for blanketing cargo and ballast tanks. US Patent 7 087 804, assigned to Chevron U.S.A. Inc. (San Ramon, CA); 2006. http://www.freepatentsonline.com/7087804.html.

Alkyl sulfides are resistant to oxidation; however, they do not have such a strong odor as the mercaptans and therefore they are used as stand-alone odorants. They are added primarily to lower the freezing point of mercaptans. The common odorant blends fall into one of the following main categories [2]:

1. mercaptan blends,
2. mercaptan-alkyl sulfide blends,
3. tetrahydrothiophene-mercaptan blends.

Thermodynamic Properties of Odorants

The effectiveness of an odorant depends on the partition coefficients and the solubility. Vapor-liquid equilibria data for sulfur compounds in liquefied natural gas are available [41, 42].

Structure-Property Relationships

The power of an odor depends on certain properties of the respective molecule. A number of structure-odor relationships concerning odor intensity and odor quality have been established [43]. Statistical methods are used to classify the various odorants.

Other Compounds

Sulfur-containing compounds are widely known as odorants used for fuel gases but have problems in that they usually generate sulfur dioxide when the fuel gases are burned. In addition, when the fuel gases are used in fuel cells, a desulfurizer must be installed to remove odorant components that would cause catalyst poisoning [11].

Attempts have been undertaken to provide sulfur-free gas odorizing compositions. These compounds include alkyl acrylates, vinyl or alkyl ethers, *n*-valeric acid, ethyl acrylate, cyclohexene, and norbornene derivatives [44]. However, these odorants have certain disadvantages. For example, acrylic ester odorants are chemically unstable. The content of cyclohexene or ethylidene norbornene must be greater than that of mercaptans [11].

On the other hand, nitrogen-containing odorants may cause the enhanced formation of nitrogen oxides. Nitrogen oxides are toxic and react with sunlight to form ozone.

Among the alcohols suitable as odorants, geosmin is a preferred compound. "Geosmin" means *earth smell.* It is a naturally occurring organic compound which is produced by microorganisms. The human nose is extremely sensitive to geosmin. Another name for geosmin is 2,6-dimethylbicyclo[4.4.0]decan-1-ol. The structures of geosmin and other sulfur-free odorants are shown in Fig. 8.3.

In addition, *trans*-2-*trans*-4-decadienal has a low perceptual threshold [11]. Mixtures of acrylates and pyrazines have been proposed as sulfur-free odorants [44]. To protect against undesired oxidation, the odorizing compositions may comprise antioxidants—for example, butylhydroxyanisole, Ionol (ie, *tert*-butylhydroxytoluene), hydroquinone monomethyl ether, and α-tocopherol. Antioxidants are shown in Fig. 8.4.

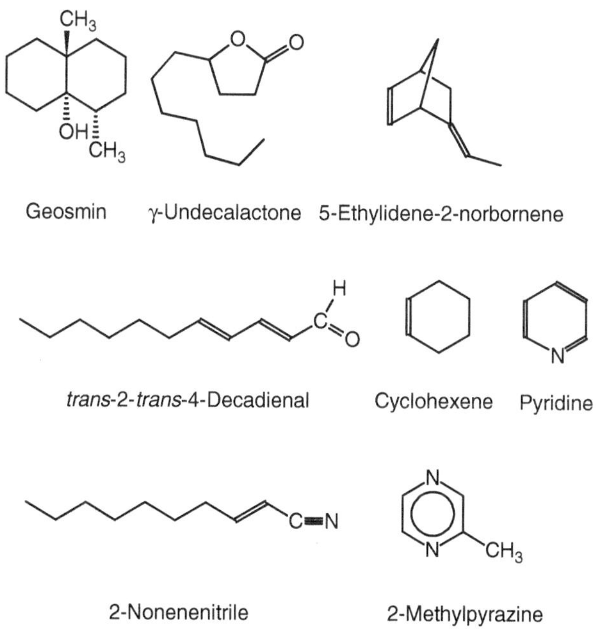

Figure 8.3 Sulfur-free odorants.

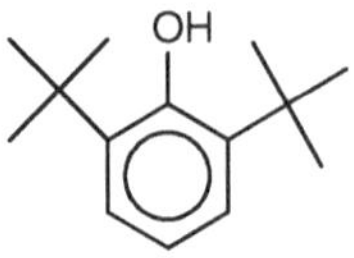

tert-Butylhydroxytoluene

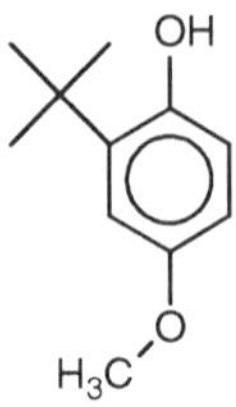

Butylhydroxyanisole

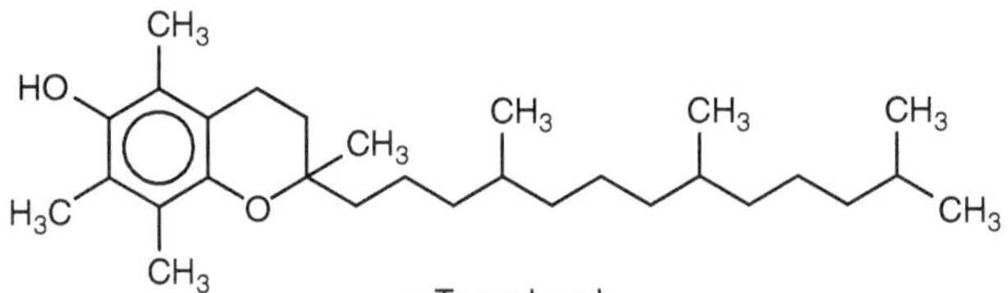

α-Tocopherol

Figure 8.4 Antioxidants.

Table 8.7 Sulfur-free additives for odorization

Additive	References
trans-2-*trans*-4-Decadienal	[11]
Geosmin	[11]
Cyclohexene	[44]
n-Valeric acid	[44]
Ethyl acrylate	[44, 45]
1-Methoxy-buten-3-yne	[46]
Methylethyl pyrazine	[46]

Sulfur-free odorants are shown in Table 8.7. Commercial sulfur-free odorants (or compositions with reduced sulfur content) include a mixture of ethyl acrylate and methyl acrylate (Gasodor S Free^{TM}) and a mixture of ethyl acrylate and tetrahydrothiophene (Spotleak Z) [47].

Sulfur-free odorants are discussed controversially. They can be readily smelled but the smell resembles garlic; therefore humans do not associate this type of smell with the typical smell of combustible gas because humans are accustomed to the mercaptan smell.

Some cities have changed their odorant in the natural gas pipelines to a sulfur-free odorant. There are reports of good success [48]. However, some cities changed back to the conventional sulfur-based odorants.

The behavior of Gasodor S Free^{TM} during the reforming of methane that has been odorized with just this odorant has been tested with respect to the application of methane in fuel gas systems. It has been verified that the odorization with Gasodor S Free™ would not have a negative impact on the subsequent use in fuel gas systems [49].

8.1.4 Industrial Synthesis of Odorants

The chemistry of thiols has been reviewed by Roberts [50]. Ethyl mercaptan is readily formed by the reaction of ethyl bromide with hydrogen sulfide. Further, mercaptans can be produced by the reaction of hydrogen sulfide and an olefin in the presence of a catalyst. Ethyl mercaptan can be prepared by the reaction of pure ethylene and hydrogen sulfide without creating a problem for separation because the desired mercaptan is the only mercaptan product. The basic reactions are:

$$CH_3 - CH_2 - Br + H_2S \rightarrow CH_3 - CH_2 - SH + HBr \qquad \text{(ii)}$$

and

$$CH_2 = CH_2 + H_2S \rightarrow CH_3 - CH_2 - SH \qquad \text{(iii)}$$

However, where the reaction mixture comprises more than one olefin, such as ethylene and propylene, it is likely to produce both ethyl mercaptan and propyl mercaptans, resulting in separation difficulty, because these mercaptans cannot be easily separated by distillation, extraction, filtration, and membrane diffusion. However, the selective production of ethyl mercaptan from a fuel gas mixture can be achieved by special catalysts (ie, in the presence of oxides of cobalt and molybdenum) [51].

Mercaptans and sulfides can be selectively produced from alcohols. The alcohol is reacted with hydrogen sulfide in the presence of a catalyst blend. The alcohols can include primary and secondary alcohols [52]. The reaction is of the electrophilic substitution type:

$$R - OH + H_2S \rightarrow R - SH + H_2O \qquad \text{(iv)}$$

In addition, mercaptans are formed by the reaction of sulfides with hydrogen sulfide:

$$R - S - R + H_2S \rightarrow 2R - SH \qquad \text{(v)}$$

8.1.5 Uses and Properties

Odorant Injection Techniques

Odorants are typically provided in liquid form and are added to the gas at a location where distribution gas is taken from a main gas pipeline and provided to a distribution pipeline. In such circumstances, the gas pressure may be stepped down through a regulator to a lower pressure. The odorants which are added to natural gas are extremely concentrated. Odorants such as *tert*-butyl mercaptan and other blends are mildly corrosive and are also very noxious.

If a leak of odorant occurs at an injection site, people in the surrounding area will assume that a gas leak has occurred, with areas being evacuated and commerce being interrupted. Contrarily, if such mistakes become common, people in the surrounding area will become desensitized to the smell of a potential gas leak and will fail to report legitimate leaks.

Two techniques are commonly used to odorize natural gas in a main distribution pipeline. The first technique involves the injection of liquid odorant directly into the pipeline. A high-pressure injection pump pumps the odorant from a liquid storage tank into a small pipe which empties directly into the main gas pipeline. Because the odorant is extremely volatile, drops injected into the pipeline immediately disperse and spread throughout the gas in the pipeline. In this way, within a few seconds, the drops of liquid odorant are dispersed in gaseous form [53].

The flow of gas in the pipeline is typically metered so that liquid odorant can be injected periodically. For example, a few drops of odorant will suffice for a 30-m^3 flow of natural gas. When the gas flow meter indicates that such an amount of natural gas has flowed through the pipe, the corresponding previously determined amount of liquid odorant is injected into the pipeline. As another indicated amount of gas flows past the injection site, another injection of odorant is performed.

Even though the injection is performed on a periodic basis, the odorant diffusion within the gas provides adequate levels of odorant throughout the pipeline, assuming the time between injections is not too great.

Another technique for odorizing a natural gas pipeline involves bypassing a small amount of natural gas at a slightly higher pressure than the pressure of the main distribution pipeline through a tank containing a liquid odorant. This bypass gas absorbs relatively high concentrations of the odorant while it is in the tank. This heavily odorized bypass gas is then placed back into the main pipeline and the odorant, now volatalized, diffuses throughout the pipeline in much the same manner as described with respect to the liquid injection system [54].

Another method for odorizing natural gas flowing through a pipeline is the injection of the odorant into the pipeline at a controlled rate. The system includes an odorant storage tank containing the odorant to be injected. A pressurized source of inert gas, such as nitrogen, communicates with the odorant storage tank to maintain the tank at a desired positive pressure above the pressure of the natural gas pipeline. An injection conduit connects the odorant storage tank with the pipeline. A photooptic metering system, located within the injection conduit, meters the odorant to be injected into the pipeline [55].

In a still further improved technique, the chemical is metered on a dropwise basis, with individual drops of the chemical being counted as they pass through a measuring unit into the injection conduit and into the pipeline. The measuring unit includes ultrasonic transmitters and receivers, which act as either proximity sensors or by measuring the transit time to provide a measurement of the flow rate of the odorant either on a drop basis or in a steady-state flow condition.

Alternatively, liquid drops land on the diaphragm of a piezoelectric sensor and thereby generate sound waves. The sound waves are transmitted to an associated crystal, which, in turn, generates a proportional electric charge, which results in a voltage difference between two electrodes. The resulting voltage spikes can be counted and measured [56].

Leak Detection

Leaks in pipelines can be detected by means of a test fluid. The test fluid, a mixture of dimethyl sulfide in solvent, is injected into a pipeline. In the case of a leak, the test fluid escapes through the leak, and the odorant is released from the closed compartments [57].

Fuel Cells

With the advent of fuels cells based on hydrogen, odorization has become an issue in the odorization of hydrogen [12]. Here another problem emerges, as odorants for fuel cell systems may have negative impacts on the performance of the fuel cells.

Commercial odorants used in gas leak detection act as poisons for the catalysts used in hydrogen-based fuel cells, most specifically for proton exchange membrane fuel cells. Chemical compounds based on mixtures of acrylic acid and nitrogen compounds have been adopted to achieve a sulfur-free odorization of a gas [58].

In the use of natural gas and other petroleum gases for hydrogen generation for fuel cell applications, sulfur-free natural or petroleum gases are needed, or else a desulfurization step must be incorporated in the reforming process, which adds further cost to hydrogen generation.

Fuel cells are sulfur intolerant because sulfur compounds poison the noble metal catalysts used in these fuel cells. If sulfur-containing odorants are used, it will be necessary to remove sulfur-containing materials, such as mercaptan odorants, from the feed gas with use of materials such as zinc oxide. However, some sulfur-containing materials, such as thiophenes, cannot be removed by zinc oxide and may require a specific hydrodesulfurization process, using hydrogen gas, to remove sulfur.

A further complexity for hydrogen fuel comes from the nature of the hydrogen flame propagation. When gases burn in air, their flames propagate upward with greater ease than they propagate downward. This is primarily due to the natural convection of hot burned gases in an upward direction. For petroleum gases, propane and methane, the upward and downward propagating lean limits of combustion are approximately the same.

However, for hydrogen, because they differ by a factor of 2.5, the amount of odorant needed for leak detection in hydrogen could be more than 2.5 times that needed for methane or propane. The higher quantity of the odorant needed for hydrogen odor detection further complicates the sulfur poisoning problems for hydrogen gas used in fuel cells [58].

Odor Fading

One specific problem of odorization is odor fading. The gas may be satisfactorily odorized at the source, but if it no longer has the necessary odor impact and intensity by the time it reaches the customer, escaping gas can remain undetected and result in a serious fire or explosion hazard [2]. Basically three causes for fading may arise [2]:

1. Oxidation, the formation of disulfides in the presence of iron oxide and oxygen traces.
2. New pipe materials may cause adsorption or absorption of the odorant on the surface of a plastic pipe.
3. Gas quality problems may cause masking, or reaction of odorant components with impurities in the gas stream.

The presence of rust and air within a pipeline may act as a catalyst for the oxidation of mercaptans. This results in compounds that cannot be smelled at all. On the other hand, sulfide components are much more resistant to oxidation.

Dry gas is the easiest to odorize and does not cause odor fade. Condensed liquids in the pipeline may absorb components of the odorant. Odor masking may also occur owing to the odor imparted by any impurities present in the gas.

Odor fading from odorized liquefied petroleum gas stored in carbon steel containers can occur by catalytic effects of the containers. To postpone this effect, the respective steel surfaces can be deactivated by treatment of the surface with a deactivating agent [59] before the walls are exposed to the liquefied petroleum gas containing the odorizer. Examples of such deactivating agents are benzotriazole, tolyl triazole, mercaptobenzothiazole, benzothiazyl disulfide, or mixtures of these compounds [60]. It has been suggested that a mathematical model and adequate software should be developed to predict odorant fading [61].

Environmental Problems

If natural gas for storage in natural reservoirs is odorized with sulfur compounds, then a possible environmental impact can result. Some of the odorant is lost in the formation [62]. If the loss occurs in a reservoir adjacent to an aquifer, it could contaminate the water and cause environmental problems. When gas is drawn off, water is also often injected into the reservoir. A case was described in which the respective water had a strong characteristic odor [63]. A stripping column has been recommended to overcome this problem.

Contaminated groundwater can also be decontaminated by the reaction with iron [64]. This technique was proposed to remedy groundwater contaminated with ethyl mercaptan in situ. Studies suggest chemical reactions with iron rather than an irreversible surface adsorption. Gas odorizers can be removed by extraction, similarly to the usual glycol dehydration and desulfurization process [65].

Another cleaning process for the removal of tetrahydrothiophene uses an advanced oxidation technique, consisting of water treatment by ultraviolet radiation in combination with a dose of hydrogen peroxide [66]. It is possible to keep the concentrations of the odorant and condensate in the effluent below 0.1 ppb.

8.1.6 Removal of Odorants

The removal of a sulfur odorant from pipeline natural gas can prevent or delay the corrosion of gas turbines. This procedure may increase their lifetime and efficiency. The removal of the odorant *tert*-butyl mercaptan by selective adsorption with fiber sorbent modules and the removal with pellet packed beds have been compared with respect to capital and operating costs [67].

The capital costs were estimated on the basis of the size of the equipment needed for each system, which varies depending on the regeneration gas used. The operating costs were estimated on the basis of the utilities needed to run each system, again varying depending on the regeneration gas used, and also on whether heat integration is needed. It has been found that use of a fraction of the product gas for regeneration is safer, simpler, and more economical than use of compressed air or nitrogen. Because of the need for more adsorbing beds to process the same amount of *tert*-butyl mercaptan-containing pipeline natural gas, the capital cost associated with a pellet packed bed system is significantly higher than that of a fiber sorbent system. The integration of heat into the sorption system, using the energy generated from waste gas combustion to produce steam, dramatically decreases the operating cost by reducing the parasitic load. It has been established that a fiber sorbent system with heat integration can be an attractive technology in comparison with packed bed systems for the removal *tert*-butyl mercaptan [67].

REFERENCES

[1] Quaglio J. Wassergas als Brennstoff der Zukunft: Strong's Patent zur Bereitung von Heizgas in Verwendung mit Lowe's Verfahren für Leuchtgas. Bericht 252123468. Wiesbaden: Bergmann; 1880.
[2] Usher MJ. Odor fade—possible causes and remedies. In: Proceedings Volume, no. 285. CGA Gas Measurement School. Ontario, CA: Canadian Gas Association; 1999.
[3] Muhlbauer WK. Distribution systems. In: Pipeline risk management manual. 3rd ed. Burlington: Gulf Professional Publishing; 2004. p. 223–42 [chapter 11]. ISBN 978-0-75-067579-6, http://dx.doi.org/10.1016/B978-075067579-6/50014-5. http://www.sciencedirect.com/science/article/B86HJ-4P9MNX3-Y/2/050f6809931a54836eadcea87154ff28.
[4] Fant EE. Odorization—a regulatory perspective. In: Odorization, vol. 3. Chicago, IL: Institute of Gas Technology; 1993. p. 109–18.
[5] Henderson DF. Large volume odorization, installation, operation, and maintenance. In: Odorization, vol. 3, Chicago, IL: Institute of Gas Technology; 1993. p. 239–49.
[6] Toth JM. Natural gas odorization and its techniques. In: Proceedings Volume. 49th Annu. Appalachian Gas Meas Short Course (Coraopolis, Penn, 8/15–18/89); 1989. p. 170–4.
[7] Oudman P. Odorization and odorant monitoring practices at Canadian Western Natural Gas Company Limited. In: Wilson GG, editor. Odorization, vol. 3. Chicago, IL: Institute of Gas Technology; 1993. p. 389–405.
[8] Holtappels K, Brinkmann C, Dietlen S, Schröder V, Stickling J, Schönbucher A. Messung und Simulation des Inertgaseinflusses auf Explosionsgrenzen bei erhöhten Anfangsdrücken. Chem Ing Tech 2001;73(3):270–4. http://dx.doi.org/10.1002/1522-2640(200103)73:3<270::AID-CITE270>3.0.CO;2-V.
[9] Askar E, Schröder V, Acikalin A, Steinbach J. Calculation of flammability limits of gas phases with ethylene oxide in sterilisers. Biomed Eng 2008;53(6):265–9.

[10] Morley C. GasEq: a chemical equilibrium program for windows; 2005. [electronic:] http://www.gaseq.co.uk/.
[11] Kato Y. Odorant for fuel gas. US Patent 7 182 796, assigned to Soda Aromatic Co., Ltd. (JP); 2007. http://www.freepatentsonline.com/7182796.html.
[12] Kopasz JP. Fuel cells and odorants for hydrogen. Int J Hydrogen Energy 2007;32(13):2527–31. http://www.sciencedirect.com/science/article/B6V3F-4MNHXYJ-1/2/5c30edfdc59e5462b2e6c3634660f544.
[13] Klusmann EB. Odor monitoring at Southern California Gas Company. In: Odorization, vol. 3. Chicago, IL: Institute of Gas Technology; 1993. p. 407–23.
[14] Wetteman AJ, Wilson JR. Operation of large volume odorizers. In: Odorization. vol. 3. Chicago, IL: Institute of Gas Technology; 1993. p. 251–5.
[15] Patnaik P. A comprehensive guide to the hazardous properties of chemical substances. 3rd ed. New York: John Wiley & Sons; 2007.
[16] Ripley DL, Goetzinger JW, Whisman ML. Human response research evaluation of alternate odorants for LP-gas. GPA Res. Rep. RR-129; 1990.
[17] Quaife LR, Moynihan KJ. A new pipeline leak-locating technique utilizing a novel odourized test-fluid (patent pending) and trained domestic dogs. In: Proceedings Volume. 1st US Environ. Protect Agency et al Oil & Gas Explor & Prod. Waste Manage Pract Int. Symp. (New Orleans, 9/10–13/90); 1990. p. 647–57.
[18] Quaife LR, Moynihan KJ, Larson DA. A new pipeline leak-location technique utilizing a novel (patented) test-fluid and trained domestic dogs. In: Proceedings Volume. 71st Annu. Gpa Conv (Anaheim, Calif, 3/16–18/92); 1992. p. 154–61.
[19] Bissell J, Acker D, Quaife LR. Pipeline leak-location technique utilizing a novel test fluid and trained dogs. In: Proceedings Volume, no. 9. 5th Pipe Line Ind. & Pipes Pipelines Int. Pipeline Pigging & Integrity Monit Int. Conf. (Houston, 2/1–4/93); 1993.
[20] Flynn PJ, Sprague MJ. Hydrogen odorants and odorant selection method. US Patent 7 811 688, assigned to Enersol, Inc. N.A., LP (Fairfax, VA); 2010. http://www.freepatentsonline.com/7811688.html.
[21] Nagata Y. Measurement of odor threshold by triangle odor bag method. In: International symposium on odor measurement, Asian Network on Odor Measurement and Control. Tokyo, JP: Japan Association on Odor Environment; 2003. p. 118–27. [electronic:] http://www.env.go.jp/en/air/odor/measure/02_3_2.pdf.
[22] ASTM E 0679-04. Practice for determination of odor and taste thresholds by a forced-choice ascending concentration series method of limits. ASTM Standard, Book of Standards, Vol. 15.08 ASTM E 0679-04. West Conshohocken, PA: ASTM International; 2009.
[23] ASTM E 544-99(2004). Standard Practices for Referencing Suprathreshold Odor Intensity. ASTM Standard, Book of Standards, Vol. 15.08 ASTM E 544-99(2004). West Conshohocken, PA: ASTM International; 2009.
[24] Shimokawatoko T, Sumida K, Ueda H. Method and apparatus for measuring odorant concentration and oderant adding system. US Patent 5 844 124, assigned to Osaka Gas Co., Ltd. (Osaka, JP); 1998. http://www.freepatentsonline.com/5844124.html.
[25] Gao X, Fan H, Huang T, Wang X, Bao J, Li X, et al., Natural gas pipeline leak detector based on NIR diode laser absorption spectroscopy. Spectrochim Acta, Part A 2006;65(1):133–8. http://www.sciencedirect.com/science/article/B6VNG-4JJ86VF-2/2/8c78a47ea9b18f29f536c58b40d4a92e.
[26] Wallace JR, Stetter JR, Nacson S, Findlay Jr MW. Odorant analyzer system. EP Patent 0 445 927 assigned to Gas Research Institute; 1991. https://www.google.at/patents/EP0445927A2?cl=en.

[27] Gardner JW, Bartlett PN. A brief history of electronic noses. Sens Actuators, B 1994;18(1–3):210–11. http://www.sciencedirect.com/science/article/B6THH-44XN0C4-1B/2/43ddaa9c373dbc5a967a5af08031f517.
[28] Junichi K, Masayuki O, Hisamitsu A, Motoo K. Electronic nose. J Jpn Assoc Odor Env 2006;37(3):172–8. [In Japanese].
[29] McGill RA, Voiculescu I, Fedder GK. Microelectro-mechanical chemical sensor. US Patent 7 556 775, assigned to The United States of America as represented by the Secretary of the Navy (Washington, DC); 2009. http://www.freepatentsonline.com/7556775.html.
[30] Ameer Q, Adeloju SB, Polypyrrole-based electronic noses for environmental and industrial analysis. Sens Actuators, B 2005;106(2):541–52. http://www.sciencedirect.com/science/article/B6THH-4DFBWXF-2/2/7d6378aa3dc66395c7717b25a598a42b.
[31] Munoz BC, Pierce KJ, Galloway CP. Sensors with improved properties. US Patent 7 501 091, assigned to Smiths Detection Inc. (Pasadena, CA); 2009. http://www.freepatentsonline.com/7501091.html.
[32] Suri K, Annapoorni S, Sarkar AK, Tandon RP. Gas and humidity sensors based on iron oxide-polypyrrole nanocomposites. Sens Actuators, B 2002;81(2–3):277–82. http://www.sciencedirect.com/science/article/B6THH-44HYGRW-1/2/1a61226aa5ba2c415fc156d0db174472.
[33] TGS 842—for the Detection of Methane. tech. rep.. Arlington Heights, IL: Figaro USA, Inc.; 2009. [electronic:] http://www.figarosensor.com/products/842pdf.pdf.
[34] May RD. System and method for detecting water vapor within natural gas. US Patent 7 504 631, assigned to SpectraSensors, Inc. (Rancho Cucamonga, CA); 2009. http://www.freepatentsonline.com/7504631.html.
[35] Mazelli JR. Method of odorizing liquid natural gas. US Patent 4 025 315, assigned to San Diego Gas & Electric Co. (San Diego, CA) Dual Fuel Systems, Inc. (Los Angeles, CA); 1977. http://www.freepatentsonline.com/4025315.html.
[36] Arkema Inc., Philadelphia, PA, Gas Odorants. 2008. [electronic:] http://arkema-inc.com/literature/pdf/802.pdf.
[37] McClure JS. Warning agent and process for the odorization of a hydrocarbon fuel gas therewith; 1958.
[38] Oister WH. Odorized gas. US Patent 3545949 assigned to Pennwalt Corporation, 1970.
[39] Quaife LR, Szarka J, Moynihan KJ, Moir ME. Test-fluid composition and method for detecting leaks in pipelines and associated facilities. US Patent 5 167 867, assigned to Exxon Production Research Company; 1992. http://www.freepatentsonline.com/5167867.html.
[40] Yoshida T, Katz I, Warren CB, Wiener C. Odorization of combustible hydrocarbon gases. US Patent 4 487 613, assigned to International Flavors & Fragrances Inc. (New York, NY); 1984. http://www.freepatentsonline.com/4487613.html.
[41] Guilbot P, Valtz A, Richon D. Partition coefficients at infinite dilution for different sulfur compounds in various solvents. In: Proceedings Volume, 76th Annu. Gpa Conv (San Antonio, 3/10–12/97); 1997. p. 33–9.
[42] Kedzierski S. The solubility of odorants in natural gas (rozpuszczalnosc preparatow nawaniajacych w gazie ziemnym). Nafta Gaz (Pol) 1996;52(8):357–60.
[43] Chastrette M. Trends in structure-odor relationship. SAR QSAR Environ Res 1997;6(3):215–54. doi:10.1080/10629369708033253.
[44] Mansfeld G, Rohde U, Henke F, Kaesler H. Gas odorization method. US Patent 7 108 803, assigned to Symrise GmbH & Co. KG (DE); 2006. http://www.freepatentsonline.com/7108803.html.

[45] Charles P. Odorant mixture for odorless gas fuel. US Patent 8 137 419, assigned to Arkema France (Colombes, FR); 2012. http://www.freepatentsonline.com/8137419.html.
[46] Mueller U, Short JN. Gas odorant. US Patent 7 682 410, assigned to Givaudan SA (Vernier, CH); 2010. http://www.freepatentsonline.com/7682410.html.
[47] Heimlich F, Niebialek S, Schulz C. Odorization with spotleak Z. Gas und Wasserfach Gas Erdgas 2008;149(3):165.
[48] Wagner T. Erfahrungsbericht zur regionalen Zentralodorierung mit schwefelfreiem Odoriermittel. Gas und Wasserfach Gas Erdgas 2005;146(10):560–3.
[49] Hennings U, Reimert R, Behaviour of sulfur-free odorants in natural gas fed PEM fuel cell systems. Fuel Cells 2007;7(1).
[50] Roberts JS. Thiols. In: Kroschwitz JI, editor. Kirk-Othmer encyclopedia of chemical technology, vol. 24. 4th ed. New York: J. Wiley & Sons; 1997. p. 19.
[51] Sattich WE. Selective production of ethyl mercaptan. US Patent 5 352 838, assigned to Phillips Petroleum Company (Bartlesville, OK); 1994. http://www.freepatentsonline.com/5352838.html.
[52] Hasenberg DM, Refvik MD. Process and catalyst for synthesis of mercaptans and sulfides from alcohols. US Patent 7 399 893, assigned to Chevron Phillips Chemical Company LP (The Woodlands, TX); 2008. http://www.freepatentsonline.com/7399893.html.
[53] Marshall SE, Zeck MV. Chemical injection system. US Patent 6 208 913, assigned to YZ Systems, Inc. (Conroe, TX); 2001. http://www.freepatentsonline.com/6208913.html.
[54] Arnold JF. System and method for odorizing natural gas. US Patent 6 142 162, assigned to Odoreyes Technology, Inc. (Birmingham, AL); 2000. http://www.freepatentsonline.com/6142162.html.
[55] Zeck M. Optical odorization system. US Patent 7 056 360; 2006. http://www.freepatentsonline.com/7056360.html.
[56] Zeck M. Ultrasonic and sonic odorization systems. US Patent 7 389 786; 2008. http://www.freepatentsonline.com/7389786.html.
[57] Quaife LR, Szarka J, Moynihan KJ, Moir ME. Test-fluid composition and method for detecting leaks in pipelines and associated facilities. US Patent 5 049 312, assigned to Exxon Production Research Company (Houston, TX); 1991. http://www.freepatentsonline.com/5049312.html.
[58] Puri PS. Leak site odorization for gas leak detection. US Patent 7 229 831, assigned to Air Products and Chemicals, Inc. (Allentown, PA); 2007. http://www.freepatentsonline.com/7229831.html.
[59] Nevers AD. Odor-fading prevention from organosulfur-odorized liquefied petroleum gas. CA Patent 1 274 692 assigned to Pennwalt Corporation, Ashley Dwight Nevers, Atochem North America, Inc.; 1990. https://www.google.at/patents/CA1274692A1?cl=en.
[60] Nevers AD. Odor-fading prevention from organosulfur-odorized liquefied petroleum gas. US Patent 4 701 303; 1987. http://www.freepatentsonline.com/4701303.html.
[61] Altpeter Jr LL. Research recommended to develop odorant-fade model. Pipe Line Gas Ind 1997;80(2):39–40.
[62] Sasnanand S. Adsorption of tetrahydrothiophene in porous media: an experimental approach. Ph.D. thesis, New Mex Inst Mining Techn, Socorro, New Mexico; 1993.
[63] Girod JF, Leclerc JP, Muhr H, Paternotte G, Corriou JP. Removing a small quantity of tht (tetrahydrothiophene) from gas storage groundwater through air stripping and gas-phase carbon adsorption. Environ Progr 1996;15(4):277–82.

[64] Huang F, Lee R. Degradation of ethyl mercaptan in the presence of zero-valence iron. In: Proceedings Volume. 4th US DOE, Tulsa Univ, et al Petrol. Environ. Conf. (San Antonio, 9/9–12/97); 1997.
[65] Rojey A, Thomas M, Jullian S. Process for treatment of natural gas at a storage site. US Patent 5 803 953, assigned to Institut Francais du Petrole (Cedex, FR); 1998. http://www.freepatentsonline.com/5803953.html.
[66] Panneman HJ, Pot-Gerritsen RC, Kuiper-Van Loo EM, Pastoor H, Janssen-Van Rosmalen R. UV (ultraviolet)-oxidation process for water treatment at gas plant sites. In: Proceedings Volume, Pt. B. 20th Int. Gas Union World Gas Conf. (Copenhagen, Denmark, 6/10–13/97); 1997. p. 269, 271–85.
[67] Chen G, Lively RP, Jones CW, Koros WJ. Fiber adsorbents for odorant removal from pipeline grade natural gas. Ind Eng Chem Res 2014;53(17):7113–20. http://dx.doi.org/10.1021/ie500069y.

PART II

Refinery Chemicals

CHAPTER 9

Refinery and Feedstocks

9.1 PROCESSES IN A REFINERY

Historically, the first refinery opened in Ploieşti, Romania, in December 1856 with US investment [1]. Nowadays, modern petroleum refineries process as much as 127,000–143,000 m^3 day^{-1}.

Common processes and process units that occur in a refinery are summarized in Table 9.1. There are many process configurations of a refinery. An exemplary configuration has been described in the literature [2].

9.2 CLASSIFICATION OF PETROLEUM

Traditionally, crude oil for use in a refinery has been classified by its geographic origin [3]. However, this is not sufficient, because in one geographic region crude oils with different properties may be obtained—for instance, depending on the depth.

Additional issues for the classification are the American Petroleum Institute (API) gravity and the contents of nonhydrocarbon components (eg, sulfur). If the crude oil contains high levels of sulfur, the classification is sour, otherwise it is sweet. In case of a high API gravity, the petroleum is called a heavy crude oil, otherwise it is called a light crude oil.

The density of an oil is governed by the molecular weight and the type of the hydrocarbons. Long-chain hydrocarbons will make the petroleum denser. The hydrocarbons in crude oil can be divided into four classes, as shown in Table 9.2.

The proper classification of petroleum indicates the most recommendable use. For example, a light crude oil is preferable for refining of oil into gasoline Further, sweet petroleum is more desirable than sour petroleum as it causes less impact on the environment [3].

Guide to the Practical Use of Chemicals in Refineries and Pipelines
http://dx.doi.org/10.1016/B978-0-12-805412-3.00009-X

Table 9.1 Common processes and process units in a refinery

Process/process unit	Usage
Desalter	Washes out salt from the crude oil
Atmospheric distillation	Distills crude oil into fractions
Vacuum distillation	Further distills residual bottoms after atmospheric distillation
Naphtha hydrotreater	Uses hydrogen to desulfurize naphtha
Catalytic reforming	Converts the naphtha boiling range molecules into reformer product
Hydrotreater	Desulfurizes distillates
Catalytic cracker	Degrades heavier fractions
Hydrocracker	Uses hydrogen to upgrade heavier fractions
Visbreaking	Thermal cracking unit
Coking	Converts heavy residual oils into gasoline and diesel
Alkylation	Uses acid to produce high-octane components
Dimerization	Olefin dimerization
Isomerization	Converts linear molecules to branched molecules
Steam reforming	Hydrogen production for the hydrocracker
Liquefied gas storage	Vessels that store propane
Storage tanks	Store crude oil and finished products
Amine gas treatment	Hydrodesulfurization of hydrogen sulfide into sulfur
Cooling towers	Circulate cooling water
Boiler plants	Steam generation
Wastewater collection	
Dissolved air flotation	
Sludge biotreater	Convert sludge for reuse or disposal
Solvent refining	Removes unwanted aromatics from lubricating oil stock
Solvent dewaxing	Removes the heavy waxy constituents

Source: Wikipedia. Oil refinery—wikipedia, the free encyclopedia; 2015. [Online; accessed 25-October-2015]. https://en.wikipedia.org/w/index.php?title=Oilrefinery&oldid=684558314.

Table 9.2 Classes of hydrocarbons in crude oil

Compound	Amount (%)
Paraffins	15–60
Naphthenes	30–60
Aromatics	3–30
Asphaltics	6

Source: Anonymous. The classification of petroleum; 2013. http://www.petroleum.co.uk/classification.

9.3 QUALITY OF REFINERY FEEDSTOCKS

High-throughput methods that can be applied to the screening of refinery feedstocks to evaluate the coke and bitumen quality have been described. Such a process comprises [4]:

1. providing a plurality of refinery feedstocks;
2. placing a sample of each of the plurality of refinery feedstocks on a heating device
3. heating each sample under a vacuum to a temperature of 300–420°C and measuring the weight loss of the sample;
4. measuring the bitumen quality of the vacuum residues formed;
5. heating the vacuum residues to 450–600°C and measuring the weight loss of the sample;
6. measuring the coke quality of the products formed.

Typically, the plurality of refinery feedstocks will comprise at least 20 refinery feedstocks, such as at least 50 refinery feedstocks. The refinery feedstocks in the first step above may be any suitable refinery feedstocks, including crude oils, synthetic crude, biocomponents, visbreaker tars, solvent deasphalted pitches, atmospheric residues, vacuum residues, fuel oils, and blends of the above-mentioned components.

Suitable heating devices include optical devices and microoscillators, such as quartz microoscillators [5], or microthermal balances, as used for thermogravimetric measurements.

9.3.1 Bitumen

Bitumen, also called asphalt, is a noncrystalline, viscous material that can be derived from refinery processing of crude oils [4]. Bitumen has numerous uses, including road construction and roofing.

Bitumen can have a number of different qualities, such as stiffness and viscosity, and these qualities may be strongly dependent on the type of crude oil processed and the various processing steps used during its production.

A number of techniques may be used which can give information on the bitumen quality. Suitable analytical techniques include NMR spectroscopy and other spectroscopic techniques, such as near-infrared spectroscopy and Fourier transform infrared spectroscopy.

Asphaltenes are polar components of refinery feedstock that are generally soluble in aromatics and insoluble in paraffinic compounds, such as *n*-alkanes [4]. Thus, if the ratio of paraffinic compounds to aromatic compounds hang for example, if incompatible crude oils are blended together, asphaltenes may precipitate out of solution, and this can cause problems with equipment fouling.

The measurement of the asphaltene stability for crude oils is described in an ASTM standard [6]. A preferred method for measurement of the asphaltene stability of the vacuum residue is the critical solvent power of the vacuum residue [7].

During the extraction of bitumen from the oil sands of Alberta, large volumes of process water containing naphthenic acids are stored in the tailing ponds. Naphthenic acids are known to be toxic in aquatic environments. The online gas chromatographic separation of naphthenic acid mixtures before ultra high resolution mass spectrometry detection using electron and chemical ionization has been described [8]. The high mass resolution of the system enables the visualization of the data by Kendrick mass defect plots [9]. The addition of gas chromatographic separations allows the visualization of the data as unique compound class elution fingerprints.

The bitumen quality may also be measured by rheological and other techniques that correlate with the conventional empirical tests for bitumen quality. Suitable high-throughput techniques for the measurement of rheological and other physical properties have been described [10]. Other conventional methods for measuring the properties of bitumen have been described [6].

9.3.2 Coke

Coke is formed from the residual carbon in a crude oil, and is typically produced by cracking of the vacuum residue in a delayed coker process [4]. Other types of cokers are available, such as a fluid coker or a flexicoker.

Coking is a thermal process that cracks the vacuum residue into lighter products and a petroleum coke. The coke formed in the delayed coker (ie, green coke, petroleum coke, or raw petroleum coke) may be further calcined to produce calcine petroleum coke.

After calcination the coke becomes electrically conductive. This property may be used to measure the coke quality of the product. Other properties of calcined coke which may be used to determine its quality include air reactivity and CO_2 reactivity, which are generally dependent on the sulfur content, sodium and calcium content and content of metals, and the coefficient of thermal expansion. The structure of the coke can be elucidated by X-ray diffraction.

Calcine petroleum coke, depending on its quality, may be used for a number of processes, including the formation of carbon anodes for the aluminum industry and graphite electrodes for furnaces or for titanium dioxide production.

Coke can have a number of different qualities, dependent on the structure of the coke formed and on the impurities therein, and these qualities can be strongly dependent on the type of crude oil processed and the various processing steps used during its production [4].

The sulfur content of the vacuum residue after heating may be measured by laser induced breakdown spectroscopy [4].

During cooking, usually cracking reactions occur that produce gaseous and liquid products. These are responsible for the weight loss associated with these components. These gaseous and liquid components should be analyzed. The analysis methods include microanalysis, spectroscopic methods, and chromatography.

9.4 PROTECTION AGAINST IGNITION

There is a great wealth of knowledge and experience regarding the use of static dissipaters and conductivity improver additives [11].

9.5 COMPATIBILITY OF ADDITIVES

An example of a testing protocol for the compatibility of additives has been presented [12]. This is focused in particular on aviation-turbine fuels and aircraft fuel system materials.

9.6 AROMATIC COMPOUNDS

Aromatic compounds in refinery streams can be characterized with use of a dual-trap Fourier transform ion cyclotron resonance mass spectrometer [13].

In the inlet a mixture of species of widely different boiling points is vaporized. The interface maintains a large pressure gradient. This makes the analysis of petroleum distillates in refinery streams at very high mass resolution possible. Ions are generated by low-energy electron ionization and are transferred to the analyzer. The mass window may be increased stepwise to cover a mass range of several hundred units [13].

9.7 SULFUR CONTENT

The emission of sulfur-containing compounds into the atmosphere is an environmental concern for air pollution [14]. Combustion engines, especially those used in transportation, such as cars, trucks, or boats, are responsible for such emissions. Also, stationary power sources play a role.

In addition of being a direct source of air pollution, sulfur-containing compounds may also poison the catalytic surface of the exhaust after treatment devices. By reducing the sulfur content in the fuel and thus the amount exhausted, one can extend the useful life of such devices [14].

New power sources on the market, such as fuel cells, will also require fuel streams with lower levels of sulfur. Fuel cells burn hydrogen that has been reformed from various hydrocarbon fuels, such as gasoline. Sulfur would poison the active surfaces of a fuel cell and thus shorten its life. Therefore, various regulations have appeared demanding substantially lower acceptable levels of sulfur and sulfur-containing compounds present in the fuels for use in combustion engines [14].

The US Environmental Protection Agency regulations require the sulfur content of on-road fuel to meet the *ultra low sulfur specification*, specifically less than 15 ppm by mass of sulfur in the finished fuel [15].

The need for commercially available fuels to have continually lower levels of sulfur-containing compounds creates new challenges for the manufacturers of such fuels. The refining industry has used several different approaches for removal of sulfur from commercially available fuel feedstocks such as gasoline and diesel [14]. The methods most commonly used by the refinery industry for the removal of sulfur from fuels are hydrodesulfurization, Merox thiol extraction processing, and adsorption.

Unfortunately, in some cases, the remaining sulfur-containing compounds present in commercially available fuels are especially difficult to remove (eg, alkyl-substituted dibenzothiophenes). Therefore, in addition to processes used in the refinery directly, processes for removal of sulfur-containing compounds from fuel feedstocks after the refinery treatment have been developed. These processes are sometimes called *sulfur polishing* processes. These processes should be capable of producing fuels having particularly low concentrations of sulfur-containing fuels (ie, less 15 ppm). Further, the sulfur polishing technology must be applicable for use in a wide variety of environments without the use of heavy, large, complex, or expensive equipment [14]. Ideally, the pressures and temperatures should not exceed those normally experienced in a combustion engine.

The treatment of a hydrodesulfurized fuel with a solid adsorbent material has been described [16]. Examples of such solid adsorbents are silica gel, activated alumina, zeolites, supported CoMo sorbents, activated coke, and activated carbon.

The method consists in [16]:

1. hydrotreating the stream under conditions which include the presence of hydrogen, temperatures within the range of 200–400°C, and a catalyst comprising at least one group VIII metal and one group VI metal on an inorganic support;
2. passing the hydro-treated distillate stream to an adsorption zone containing a solid adsorbent capable of adsorbing β- and di-β-substituted dibenzothiophene aromatic sulfur compounds;
3. collecting the resulting substantially sulfur-free distillate stream;
4. regenerating the adsorbent by passing a liquid or vapor desorbent through the bed of adsorbent in the adsorption zone, thereby removing the β- and di-β-substituted dibenzothiophene sulfur compounds from the adsorbent;
5. passing the β- and di-β-substituted dibenzothiophene sulfur compound-containing desorbent to a distillation zone to separate these sulfur compounds from the desorbent, thereby resulting in a stream rich in β- and di-β-substituted dibenzothiophene sulfur compounds and a desorbent stream;
6. recycling the desorbent stream back to the adsorption zone.

An exhaust gas catalyst has been described that contains a sulfur trap warm-up catalyst housed within the exhaust stream and a sulfur scavenger component [17]. In addition, a catalyst for the absorption of nitrogen oxides is placed in the module. The sulfur-scavenging component comprises metallic trapping elements, including silver, aluminum, barium, cerium, cobalt, copper, or zinc, that may be applied to supporting materials such as high surface area materials such as alumina, zeolite, zirconia, or silica.

A desulfurization unit that can be mounted in automobiles has been described [18]. The unit is arranged between the fuel tank and an injector of the engine. The unit is also a combination of an adsorbent for adsorbing and concentrating sulfur-containing compounds and an oxidizing agent for eventually oxidizing the adsorbed sulfur-containing compounds.

The sulfur-containing compound removal unit includes a microorganism-immobilized carrier in which a microorganism capable of oxidizing the sulfur-containing oxide to a sulfate salt or a sulfite salt has been immobilized. Various microorganisms capable of oxidizing a typical aromatic sulfur-containing compound such as dibenzothiophene to a sulfate salt are known; an example is *Rhodococcus rhodochrous* ATCC53968. It is considered that this microorganism oxidizes dibenzothiophene to 2-hydroxybiphenyl and a sulfate salt via dibenzothiophene sulfoxide, dibenzothiophene sulfone, and 2-hydroxybiphenyl-2-sulfinic acid [18].

Also, by removal of an aromatic sulfur-containing compound before combustion, not only can the quantity of a sulfur oxide in the exhaust gas discharged from an engine be reduced, but also the quantity of particulate matter in the exhaust gas can be reduced and the durability of a catalyst for removal of nitrogen oxide can be improved. The quantity of the particulate matter in the exhaust gas is reduced by half and the durability of the catalyst for removal of nitrogen oxide is improved by a factor of about 2 [18].

Also, fuel filters that can be used with fuel cells have been described [14].

9.8 FOULING DEPOSITS

The deposition of fouling materials can be accessed by a standard published by the National Association of Corrosion Engineers [19].

A method for evaluating the fouling caused by refinery feedstocks has been described [20]. The method consists in sampling a plurality of refinery feedstocks or a fractions thereof, establishing an array of metal samples that are representative of the refinery, contacting the metal samples with each of the feedstock samples under appropriate conditions, and analyzing the fouling of the samples.

To speed up the procedure, the metal samples can be in the form of a microfabricated arrays and the testing is done in parallel by a high-throughput experimentation method. The parameters that are important for the contact of the metal samples with the feedstock samples are summarized in Table 9.3.

In the asphaltene stability test, samples of the refinery feedstock are mixed with solvents (eg, *n*-heptane, toluene) at different volumes and compositions of solvent and each is subjected to vibrational mixing. The asphaltene stability is determined by the onset of flocculation. The onset of flocculation due to the addition of the solvents may be determined

Table 9.3 Experimental parameters

Parameter	
Feedstock type	Feedstock fraction
Sample geometry	Contact time
Temperature	Pressure
Flow rate	Turbulence
Sheer	Soak

Source: Butler G, Couves JW, Greenough P, Gudde NJ, Hodges MG. Process for evaluating fouling caused by refinery feedstocks. US Patent 7984643, assigned to BP Oil International Limited (Middlesex, GB); 2011. http://www.freepatentsonline.com/7984643.html.

by infrared spectroscopy, measuring the change in the transmission of the radiation.

The method can be used to test the effectiveness of different chemical treatments (eg, different additives or different doses) to identify an optimal treatment for a particular fouling blend. Thus it is possible to assess the suitability of fouling inhibitors by the addition of such inhibitors to the refinery feedstock.

A number of different potential fouling inhibitors may be assessed against a variety of feedstock blends, enabling the optimization of the blend and the treatment. By use of metal samples coated with particular surface treatments, the ability of such treatments to reduce fouling can be determined.

9.8.1 Modeling Fouling

The concept of a thermal *fouling threshold*, first introduced by Ebert and Panchal [21], has been revisited [22]. Models of hydraulic effects of fouling to provide a graphical tool, the modified temperature field plot, for assessment of chronic chemical reaction fouling effects in refinery heat exchangers have been presented.

Fouling data of differing quality collected from a pilot plant and from a refinery operation were compared with two threshold fouling models and one fouling model based on the Epstein deposition model. The Epstein model showed the best agreement, mainly because it can accommodate fouling that is mass transfer controlled as well as reaction controlled. Hydraulic analysis indicated that the simple slab approximation for fouling layers may give a reasonably good mapping between heat transfer and pressure drop effects as long as roughness contributions are not significant.

9.9 MONITORING CATALYST REQUIREMENTS

A method for monitoring a catalyst injection system or a fluid catalytic cracking unit has been described [23]. The method includes automatically updating catalyst available inventory information in a digital memory device in response to a catalyst use event [23].

The exhaust system of the unit is coupled to the fluid catalytic cracking unit and is adapted to control and to monitor the exhausted by-products of the fluid cracking process. The controller in the system is typically housed in an explosion-proof enclosure to prevent spark ignition of gases which may potentially exist on the exterior of the enclosure in a petroleum processing environment.

The sufficiency of the updated catalyst available is automatically determined. A resupply action is then taken in response to a determination of insufficient catalyst available inventory. The method is repeated for each catalyst use event [23].

9.10 HEAVY HYDROCARBON CONTAMINATION

There are numerous, industry-accepted test methods for the analysis of physical and chemical properties of hydrocarbons which have been used to detect the presence of hydrocarbon contaminants in various refinery process streams and also a to classify them. These methods are summarized in Table 9.4.

Conventionally, these tests are conducted with many different types of apparatuses. These tests are time-consuming and relatively expensive [42]. These tests also lack sensitivity to detect contaminants at low concentrations and lack specificity to determine the type of contaminant. The contamination of light refinery process streams with heavy aromatics is an ongoing concern in the petroleum industry.

Such contamination can cause severe processing problems or result in final products which do not meet contractual or government specifications. To mitigate and control this problem in particular, it is desired to detect the presence and type of the heavy aromatics and to identify the source. Often, the source of contamination is one or more heat exchanger leaks occurring in the system [42].

Table 9.4 Test methods for hydrocarbons

Standard	Description	Reference
ASTM D86	Distillation temperatures at various percentages distilled	[24]
ASTM D4052	Density and API gravity	[25]
ASTM D1298	Test for API gravity	[26]
ASTM D4629	Nitrogen traces	[27]
ASTM D4808	Hydrogen content	[28]
ASTM D2622	Sulfur content	[29]
ASTM D2887	Boiling range distribution	[30]
ASTM D3710-95	Boiling range distribution	[31]
ASTM D2789-25	Hydrocarbon types in low olefinic gasoline by mass spectrometry	[32]
ASTM D4534	Benzene content of cyclic products	[33]
ASTM D5134	Detailed analysis of petroleum naphtha through *n*-nonane by capillary gas chromatography	[34]
ASTM D6293-98	Oxygenates and paraffin, olefin, naphthene, aromatic fuels by gas chromatography	[35]
ASTM D6296-98	Total olefins by multidimensional gas chromatography	[36]
ASTM D5443-14	Paraffin, naphthene, and aromatic hydrocarbon type analysis in petroleum distillates	[37]
ASTM D5769-10	Determination of benzene, toluene, and total aromatics in finished gasoline	[38]
ASTM D893	Pentane insolubles	[39]
ASTM D445	Kinematic viscosity	[40]
ASTM D93	Flash point	[41]

API, American Petroleum Institute.
Source: Triche NR, Lubeck AJ. Detection and classification of heavy hydrocarbon contamination in refinery process streams via spectrofluorometry. US Patent 7 939 335, assigned to Marathon Ashland Petroleum LLC (Findlay, OH); 2011. http://www.freepatentsonline.com/7939335.html.

An improved method for the detection of heavy hydrocarbon components in at least one sample taken from a distillate stream or reformer feed contains the following steps [42, 43]:

1. measuring the fluorescence by spectrofluorometry for determination of the concentration of a heavy hydrocarbon component;
2. transforming the fluorescence data into the concentration of the respective hydrocarbon component;
3. outputting the signal to a computer and analysis of the data.

Spectrofluorometry is a chemical analysis method in which a sample exposed to light at one wavelength absorbs this radiation and emits light at another, longer wavelength. The intensity of the emission is proportional to the concentration of the fluorescing species. For samples containing more than one fluorescing species, the excitation and emission occur not at a single wavelength but over a range of wavelengths.

Luminescence occurs when a material is activated by an energy stimulus whereby the molecules are raised to an excited state. When the stimulus ends, these molecules drop back to a less excited state and give off light energy while doing so. If the emitted light energy lasts for a short time, the effect is referred to as fluorescence and is referred to as phosphorescence when the emitted energy lasts for a longer duration.

The sample of material excited with light of a particular wavelength emits light energy in the form of an emission spectrum whose amplitude profile, over the range of wavelengths emitted, constitutes a fingerprint which gives the identity and nature of the unknown material in the sample.

For example, the optimum excitation and emission wavelengths increase with the molecular size of aromatic compounds. Most aromatic hydrocarbons are strongly fluorescing species such that the spectrofluorometric method provides a very sensitive method for detecting heavy hydrocarbon contamination in a reformer feed (ie, naphtha).

In a virgin naphtha, the aromatics are predominantly mononuclear, containing only a single benzene ring. In contrast, virgin diesel has a predominance of two-ring aromatics and naphthalenes with some three-ring structures. Gas oils consist mostly of three-ring or larger aromatics. The spectrofluorometric method measures the differences in the aromatics among these streams and so allows one to detect traces of heavy hydrocarbons [42].

REFERENCES

[1] Popa L. Romania. In: Perks D, editor. Oil & gas: a comparative guide to the regulation of oil and gas projects: jurisdictional comparisons. London: European Lawyer Reference Series; 2012. p. 167. ISBN 0414027078.

[2] Wikipedia. Oil refinery—wikipedia, the free encyclopedia; 2015. [Online; accessed 25-October-2015], https://en.wikipedia.org/w/index.php?title=Oil_refinery&oldid=684558314.

[3] Anonymous. The classification of petroleum; 2013. http://www.petroleum.co.uk/classification.

[4] Hodges MG. Process for evaluating quality of coke and bitumen of refinery feedstocks. US Patent 7 803 627, assigned to BP Oil International Limited (Middlesex, GB); 2010. http://www.freepatentsonline.com/7803627.html.

[5] Spates JJ, Martin SJ, Mansure AJ. Acoustic-wave sensor apparatus for analyzing a petroleum-based composition and sensing solidification of constituents therein. US Patent 5 661 233, assigned to Sandia Corporation (Albuquerque, NM); 1997. http://www.freepatentsonline.com/5661233.html.

[6] ASTM D7060-12. Standard Test Method for Determination of the Maximum Flocculation Ratio and Peptizing Power in Residual and Heavy Fuel Oils (Optical Detection Method). ASTM Standard, Book of Standards, Vol. 05.04 ASTM D7060-12. West Conshohocken, PA: ASTM International; 2012.

[7] Kimbrell MR, Nemana S, Zaluzec E. Predictive crude oil compatibility model. WO Patent 2004 061 450 assigned to Bp Corp North America Inc.; 2004. https://www.google.at/patents/WO2004061450A1?cl=en.

[8] Ortiz X, Jobst KJ, Reiner EJ, Backus SM, Peru KM, McMartin DW, et al. Characterization of naphthenic acids by gas chromatography-Fourier transform ion cyclotron resonance mass spectrometry. Anal Chem 2014;86(15):7666–73. http://dx.doi.org/10.1021/ac501549p.

[9] Kendrick E. A mass scale based on CH_2 = 14.0000 for high resolution mass spectrometry of organic compounds. Anal Chem 1963;35(13):2146–54. http://dx.doi.org/10.1021/ac60206a048.

[10] Hajduk DA, Kolosov O. High throughput preparation and analysis of plastically shaped material samples. US Patent 7 013 709, assigned to Symyx Technologies, Inc. (Santa Clara, CA); 2006. http://www.freepatentsonline.com/7013709.html.

[11] ASTM D4865-09. Standard guide for generation and dissipation of static electricity in petroleum fuel systems. ASTM Standard, Book of Standards, Vol. 05.02 ASTM D4865-09. West Conshohocken, PA: ASTM International; 2009.

[12] ASTM D4054-09. Standard practice for qualification and approval of new aviation turbine fuels and fuel additives. ASTM Standard, Book of Standards, Vol. 05.02 ASTM D4054-09. West Conshohocken, PA: ASTM International; 2009.

[13] Guan S, Marshall AG, Scheppele SE. Resolution and chemical formula identification of aromatic hydrocarbons and aromatic compounds containing sulfur, nitrogen, or oxygen in petroleum distillates and refinery streams. Anal Chem 1996;68(1):46–71. http://dx.doi.org/10.1021/ac9507855.

[14] Rohrbach RP, Zulauf GB, Unger PD, Bause DE. Apparatus and method for removing sulfur containing compounds from a post-refinery fuel stream. US Patent 7 575 688, assigned to Honeywell International Inc. (Morristown, NJ); 2009. http://www.freepatentsonline.com/7575688.html.

[15] Burgazli CR, Henry Jr CP. Synergistic additive composition for petroleum fuels. US Patent 8 821 594, assigned to Innospec Fuel Specialities LLC (Newark, DE); 2014. http://www.freepatentsonline.com/8821594.html.

[16] Savage DW, Kaul BK, Dupre GD, O'Bara JT, Wales WE, Ho TC. Deep desulfurization of distillate fuels. US Patent 5 454 933, assigned to Exxon Research and Engineering Company (Florham Park, NJ); 1995. http://www.freepatentsonline.com/5454933.html.

[17] Dou D, Molinier M, Bailey O. Sulfur trap in NO_x adsorber systems for enhanced sulfur resistance. WO Patent 2001 056 686 assigned to Delphi Tech Inc; 2002. https://www.google.at/patents/WO2001056686A9?cl=en.

[18] Sakai T, Mizuno T, Suzuki H. Apparatus for removing sulfur-containing component in fuel. US Patent 6 756 022, assigned to Toyota Jidosha Kabushiki Kaisha (Toyota, JP); 2004. http://www.freepatentsonline.com/6756022.html.

[19] NACE TM0199-2013. Standard test method for measuring deposit mass loading (deposit weight density) values for boiler tubes by the glass-bead-blasting technique. NACE Standard NACE TM0199-2013. Nashville, TN: National Association of Corrosion Engineers; 2013.

[20] Butler G, Couves JW, Greenough P, Gudde NJ, Hodges MG. Process for evaluating fouling caused by refinery feedstocks. US Patent 7 984 643, assigned to BP Oil International Limited (Middlesex, GB); 2011. http://www.freepatentsonline.com/7984643.html.

[21] Panchal CB, Kuru WC, Liao CF, Ebert WA, Palen JW. Threshold conditions for crude oil fouling. In: Bott TR, Melo LF, Panchal CB, Somerscales EFC, editors. Proc. Int. Conf. on Understanding heat exchanger fouling and its mitigation, Castelvecchio Pascoli, Italy. New York: Begell House; 1999. p. 273–82.

[22] Yeap BL, Wilson DI, Polley GT, Pugh SJ. Mitigation of crude oil refinery heat exchanger fouling through retrofits based on thermo-hydraulic fouling models. Chem Eng Res Design 2004;82(1):53–71. http://www.sciencedirect.com/science/article/pii/S0263876204724594, http://dx.doi.org/10.1205/026387604772803070.

[23] Evans M. Method for monitoring catalyst requirements of a refinery. US Patent 8 099 259, assigned to Intercat Equipment, Inc. (Sea Girt, NJ); 2012. http://www.freepatentsonline.com/8099259.html.

[24] ASTM D86-12. Standard Test Method for Distillation of Petroleum Products at Atmospheric Pressure. ASTM Standard, Book of Standards, Vol. 05.01 ASTM D86-12. West Conshohocken, PA: ASTM International; 2012.

[25] ASTM D4052. Standard Test Method for Density, Relative Density, and API Gravity of Liquids by Digital Density Meter. ASTM Standard, Book of Standards, Vol. 05.02 ASTM D4052. West Conshohocken, PA: ASTM International; 2011. http://dx.doi.org/10.1520/D4052-11.

[26] ASTM D1298-12b. Standard test method for density, relative density, or API gravity of crude petroleum and liquid petroleum products by hydrometer method. ASTM Standard, Book of Standards, Vol. 05.01 ASTM D1298-12b. West Conshohocken, PA: ASTM International; 2012.

[27] ASTM D4629. Standard test method for trace nitrogen in liquid petroleum hydrocarbons by syringe/inlet oxidative combustion and chemiluminescence detection. ASTM Standard, Book of Standards, Vol. 05.02 ASTM D4629. West Conshohocken, PA: ASTM International; 2012. http://dx.doi.org/10.1520/D4629-12.

[28] ASTM D4808. Standard test methods for hydrogen content of light distillates, middle distillates, gas oils, and residua by low-resolution nuclear magnetic resonance spectroscopy. ASTM Standard, Book of Standards, Vol. 05.01 ASTM D4808. West Conshohocken, PA: ASTM International; 2012. http://dx.doi.org/10.1520/D4808-01R12.

[29] ASTM D2622-10. Standard test method for sulfur in petroleum products by wavelength dispersive X-ray fluorescence spectrometry. ASTM Standard, Book of Standards, Vol. 05.01 ASTM D2622-10. West Conshohocken, PA: ASTM International; 2010.

[30] ASTM D2887. Standard test method for boiling range distribution of petroleum fractions by gas chromatography. ASTM Standard, Book of Standards, Vol. 05.01 ASTM D2887. West Conshohocken, PA: ASTM International; 2013. http://dx.doi.org/10.1520/D2887.

[31] ASTM D3710-95. Standard test method for boiling range distribution of gasoline and gasoline fractions by gas chromatography. ASTM Standard, Book of Standards, Vol. 05.01 ASTM D3710-95. West Conshohocken, PA: ASTM International; 2004.

[32] ASTM D2789-25. Standard test method for hydrocarbon types in low olefinic gasoline by mass spectrometry. ASTM Standard, Book of Standards, Vol. 05.01 ASTM D2789-25. West Conshohocken, PA: ASTM International; 2011.

[33] ASTM D4534. Standard test method for benzene content of cyclic products by gas chromatography. ASTM Standard, Book of Standards, Vol. 05.01 ASTM D4534. West Conshohocken, PA: ASTM International; 1999 [Withdrawn].

[34] ASTM D4534. Standard test method for detailed analysis of petroleum naphthas through *n*-nonane by capillary gas chromatography. ASTM Standard, Book of Standards, Vol. 05.02 ASTM D4534. West Conshohocken, PA: ASTM International; 2013.
[35] ASTM D6293-98. Standard test method for oxygenates and paraffin, olefin, naphthene, aromatic (O-PONA) hydrocarbon types in low-olefin spark ignition engine fuels by gas chromatography. ASTM Standard ASTM D6293-98. West Conshohocken, PA: ASTM International; 2003 [Withdrawn].
[36] ASTM D6296-98. Standard test method for total olefins in spark-ignition engine fuels by multidimensional gas chromatography. ASTM Standard, Book of Standards, Vol. 05.03 ASTM D6296-98. West Conshohocken, PA: ASTM International; 2013.
[37] ASTM D5443-14. Standard test method for paraffin, naphthene, and aromatic hydrocarbon type analysis in petroleum distillates through 200C by multi-dimensional gas chromatography. ASTM Standard, Book of Standards, Vol. 05.02 ASTM D5443-14. West Conshohocken, PA: ASTM International; 2014.
[38] ASTM D5769-10. Standard test method for determination of benzene, toluene, and total aromatics in finished gasolines by gas chromatography/mass spectrometry. ASTM Standard, Book of Standards, Vol. 05.02 ASTM D5769-10. West Conshohocken, PA: ASTM International; 2010.
[39] ASTM D893. Standard test method for insolubles in used lubricating oils. ASTM Standard, Book of Standards, Vol. 05.01 ASTM D893. West Conshohocken, PA: ASTM International; 2014. http://dx.doi.org/10.1520/D0893.
[40] ASTM D445. Standard test method for kinematic viscosity of transparent and opaque liquids (and calculation of dynamic viscosity). ASTM Standard, Book of Standards, Vol. 05.01 ASTM D445 West Conshohocken, PA: ASTM International; 2014. http://dx.doi.org/10.1520/D0445-14E02.
[41] ASTM D93. Standard test methods for flash point by Pensky-Martens closed cup tester. ASTM Standard, Book of Standards, Vol. 05.01 ASTM D93. West Conshohocken, PA: ASTM International; 2013. http://dx.doi.org/10.1520/D0093.
[42] Triche NR, Lubeck AJ. Detection and classification of heavy hydrocarbon contamination in refinery process streams via spectrofluorometry. US Patent 7 939 335, assigned to Marathon Ashland Petroleum LLC (Findlay, OH); 2011. http://www.freepatentsonline.com/7939335.html.
[43] Triche NR, Lubeck AJ. Detection and classification of heavy hydrocarbon contamination in refinery process streams via spectrofluorometry. US Patent 7 501 285, assigned to Marathon Ashland Petroleum LLC (Findlay, OH); 2009. http://www.freepatentsonline.com/7501285.html.

CHAPTER 10

Special Refinery Additives

10.1 ACETYLENIC SURFACTANTS

A class of acetylenic surfactants can be used to resolve or break water and oil emulsions. The surfactants are of particular advantage in resolving crude oil emulsions of the type encountered in desalter and similar apparatus designed to extract brines from the crude as they partition to the aqueous phase in the desalter. Also, the surfactants may be used to separate oil from oil sands and similar oil/solids matrices [1].

Crude oil contains impurities which contribute to corrosion, heat exchanger fouling, furnace coking, catalyst deactivation, and product degradation in refinery processes. These contaminants are broadly classified as [1]:

- salts,
- bottom sediment water,
- solids,
- metals.

The amounts of these impurities vary depending on the particular crude. Generally, the crude oil salt content ranges between 3 and 200 lb per 1000 barrels.

Native water present in crude oils contains predominately sodium chloride and lesser amounts of magnesium chloride and calcium chloride. On thermal hydrolysis, chloride salts are the source of highly corrosive HCl, which is severely damaging to refinery tower trays and other equipment. Additionally, carbonate and sulfate salts may be present in the crude in sufficient quantities to promote crude pre-heat exchanger scaling.

Solids other than salts are equally harmful. For example, sand, clay, volcanic ash, drilling muds, rust, iron sulfide, metal, and scale may be present and can cause fouling, plugging, abrasion, erosion, and residual product contamination. As a contributor to waste and pollution, sediment stabilizes emulsions in the form of oil-wetted solids and can carry significant quantities of oil into the waste recovery systems [1].

Guide to the Practical Use of Chemicals in Refineries and Pipelines
http://dx.doi.org/10.1016/B978-0-12-805412-3.00010-6

Metals in crude may be inorganic or organometallic compounds which consist of hydrocarbon combinations with arsenic, vanadium, nickel, copper, iron, and other metals. These materials promote fouling and can cause catalyst poisoning in subsequent refinery processes, such as catalytic cracking methods, and they may also contaminate finished products.

Most of the metals are carried as bottoms in refinery processes. When the bottoms are fed, for example, to coker units, contamination of the end-product coke is most undesirable. For example, in the production of high-grade electrodes from coke, iron contamination of the coke can lead to electrode degradation and failure in processes such as those used in the chlor-alkali industry.

Desalting is, as the name implies, a process that is adapted to remove primarily inorganic salts from the crude before refining. The desalting step is provided by addition and mixing or emulsification with the crude of a few volume percentages of freshwater to contact the brine and salt. In crude oil desalting, a water-in-oil emulsion is intentionally formed, with the water admitted being on the order of about 3–10% by volume based on the crude oil [1].

Water is added to the crude and mixed intimately to transfer impurities in the crude to the water phase. Separation of the phases occurs by the coalescence of the small water droplets into progressively larger droplets and eventual gravity separation of the oil and underlying water phase.

Demulsification agents are added, usually upstream of the desalter, and have a variety of purposes, such as to help in providing maximum mixing of the oil and water phases, dehydrate the crude oil, provide faster water separation, better salt extraction or improved solids extraction, and generate oil-free effluent water. Known demulsifying agents include water-soluble organic salts, sulfonated glycerides, sulfonated oils, acetylated castor oils, ethoxylated phenol formaldehyde resins, polyols, polyalkylene oxides, ethoxylated amines, a variety of polyester materials, and many other commercially available compounds.

Desalters are also commonly provided with electrodes to impart an electric field in the desalter. This serves to polarize the dispersed water molecules. The so-formed polarized molecules exert an attractive force between oppositely charged poles, with the increased attractive force increasing the speed of water droplet coalescence by 10- to 100-fold. The water droplets also move quickly in the electric field, thus promoting random collisions that further enhance coalescence [1].

On separation of the phases from the water-in-oil emulsions, the crude is commonly drawn off the top of the desalter and sent to the fractionator tower in crude units or other refinery processes. The water phase may be passed through heat exchangers or the like and is ultimately discharged as effluent.

In addition to the need for effective emulsion breakers in resolving the water-in-oil emulsions in desalters, water-in-oil emulsions are also commonly used in certain bitumen demulsification processes. The emulsions encountered can be of the oil-in-water type, wherein the density of the hydrocarbon materials is greater than that of water. In these cases, the hydrocarbon phase can be taken from the bottom of the vessel used for separation.

Emulsions are also formed during the production of crude oil. Water is associated with the geological formation and will be co-produced from the oil well. Also, water or steam may be added to the formation in enhanced oil recovery operations and will contribute water to the oil stream produced. Turbulence applied by choke points in the wellhead or production adds sufficient mechanical force to create an emulsion from the oil-water mixture. This water needs to be separated from the oil produced, as pipeline and other collection or transportation systems have specifications regarding the maximum amounts of water that can be associated with the oil. The water can lead to corrosion issues in the pipeline. Emulsion breakers are applied to speed the separation of the oil and water during production. Various types of equipment have been used to effect this separation, such as dehydrators and heat treaters [1].

Emulsions that become difficult to break or resolve as a result of refinery reworks, tank washes, and interfaces, are often referred to as *slop*. This slop cannot be discharged directly because of environmental concerns so it has therefore become important to efficiently resolve or separate the emulsion constituents into an oleaginous (oil) or organic phase and a combined mud and nonoleaginous water phase. The oil phase may be used as a process fluid for refinery or other processes or may be recycled for down hole use. The mud-water phase may be sent to further separation processes to separate the water for discharge or other uses and the mud for possible recycling into down hole operations. Additionally, in some cases, the drilling mud actually seeps out of the formation into the crude oil that is being extracted to form an undesirable drilling mud emulsion containing crude oil, water, and sometimes clay as components.

So, there is a need to provide effective demulsifying treatments to resolve or break water and oil emulsions, particularly the crude oil emulsions encountered in desalter apparatuses, water and bitumen emulsions, and drilling mud emulsions. The emulsions may also be encountered in heat treaters, free water knockout apparatus, inclined plate separation apparatus, water separation apparatus, hydrocyclones, and centrifuges [1].

Commercially available products are summarized in Table 10.1. The basic acetylenic diols are shown in Fig. 10.1.

The tertiary acetylenic diols, preferably 2,4,7,9-tetramethyl-5-decyne-4,7-diol and 2,5,8,11-tetramethyl-6-dodecyne-5,8-diol (see Table 10.1) are

Table 10.1 Acetylene-based surfactants

Compound
TMDD-5
TMDD-6
(TMDD-5)–1.3 mol ethoxylate
(TMDD-5)–3.5 mol ethyoxylate
(TMDD-5)–5.1 mol ethoxylate
(TMDD-5)–10.0 mol ethoxylate
(TMDD-5)–30.0 mol ethoxylate
(TMDD-6)–4.0 mol ethyoxylate
(TMDD-5)–5 mol ethoxylate/2 mol propoxylate

TMDD-5, 2,4,7,9-tetramethyl-5-decyne-4,7-diol; *TMDD-6*, 2,5,8,11-Tetramethyl-6-dodecyne-5,8-diol.
Source: Engel DB, Goliaszewski AE, McDaniel CR. Separatory and emulsion breaking processes. US Patent 7 771 588, assigned to General Electric Company (Schenectady, NY); 2010. http://www.freepatentsonline.com/7771588.html.

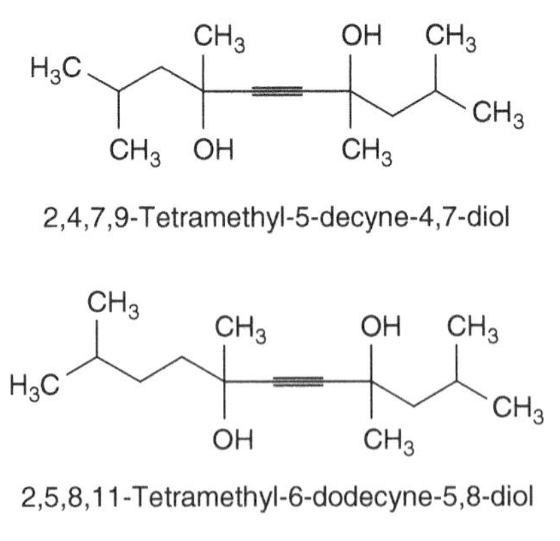

Figure 10.1 Acetylenic diols.

used as the precursors to form the ethylene oxide, propylene oxide, or mixed adducts as described in [2, 3].

As previously mentioned, the ethylene oxide and ethylene oxide/propylene oxide derivatives are also commercially available. Briefly, the procedures reported in these patents in [2, 3] involve reaction of the precursor with the requisite quantities of ethylene oxide and/or ethylene oxide followed by propylene oxide in the presence of a trialkylamine or Lewis acid.

As catalysts, trialkylamines and Lewis acids, such as BF_3, can be used. In addition, the compositions may be prepared by the reaction of a preformed acetylenic diol ethoxylate with propylene oxide in the presence of a catalyst [1].

Also, related aromatic compounds can be synthesized. For example, 2,4-dimethyl-7-phenyl-5-octyne-4,7-diol can be prepared as follows [1]: To a solution of 12.6 g 3,4-dimethyl-1-hexyn-3-ol in 500 mL diethyl ether at 0°C is added dropwise a solution of 0.22 mol *n*-butylithium in 110ml diethyl ether during a period of 1 h. The reaction mixture is then stirred for 30 min, treated with a solution of 12 g acetophenone in 100 mL ether, and allowed to warm to room temperature. The solution is then quenched with 600 mL of a 0.1 N HCl solution, and the organic phases separate. The aqueous phase is further extracted several times with ether, and the combined organic phases are washed with saturated $NaHCO_3$ solution and water and eventually dried over molecular sieves.

Also, hydroxylated actylenic compounds may be used to improve the separation of the oil from the oil sands [1].

10.2 GELLING AGENTS

The problem of tray plugging in a refinery due to residual oil gellant chemicals has been discussed in the literature [4].

The results of field testing have been presented with regard to broad field application. The objectives of this research were to optimize cost and performance, investigate any remaining questions, and establish quality control specifications based on both performance testing and NMR compositional analysis.

One issue was the ability of phosphonate esters to control volatile phosphorus at high temperatures. The distillations used to evaluate volatile phosphorus to date have had a 250°C end point. This temperature was chosen because it represents the approximate temperature experienced at the distillation tower trays where plugging has been observed from components condensing from the gas phase. However, the actual peak temperature at the tower bottom is closer to 350°C. This higher temperature is the actual temperature at which decomposition or volatilization will occur. Therefore to more fully understand our ability to control volatile phosphorus, distillations were conducted with a 350°C end point. Volatile and total phosphorus to 250 and 350°C end points have been reported.

Another continuing area of concern has been organic halide formation under distillation tower conditions. Although no organic halides were detected in either of the two initial field trials, further testing was conducted during the additional field trials reported in [4].

Also, the role of tray plugging of the refinery distillation trays from the decomposition products or volatile products of a phosphoric acid ester gelling additive has been described.

Distillation trays in the temperature range of 230–290°C produce most of the volatile phosphorus compounds that condense out of the gas phase and cause the plugging of the holes in the trays.

Plugging of trays reduces the efficiency of distillation towers. If the plugging becomes severe enough, it becomes necessary to shut down the refinery to clean the towers. Unscheduled refinery shutdowns can be extremely costly to refiners. It is also possible that phosphorus may still be present in some distillate streams such as the jet fuel draw [5].

Improved methods for fracturing subterranean formations using gelled liquid hydrocarbons have been described [6, 7]. These methods comprise the following steps: An improved gelled liquid hydrocarbon fracturing fluid is prepared comprising a hydrocarbon liquid, a gelling agent comprising a ferric iron or aluminum polyvalent metal salt of a phosphonic acid ester, a proppant material, water, and an amount of a delayed gel breaker effective to break the gelled hydrocarbon fracturing fluid.

10.3 CONDUCTIVITY ADDITIVES

Electrostatic charges can be frictionally transferred between two dissimilar, nonconductive materials. When this occurs, the electrostatic charge thus created appears at the surfaces of the contacting materials. The magnitude of the generated charge is dependent on the respective conductivity of each material. The potential for electrostatic ignition and explosion is probably greatest during product handling, transfer, and transportation [8].

Electrostatic charging is known to occur during solvent or fuel pumping operations. In such operations, the flow of low-conductivity liquid through conduits with high surface area or through fine filters combined with the disintegration of a liquid column and splashing during high-speed tank loading can result in static charging. Such static charging can result in a spark with a catastrophic potential in highly flammable environments.

Thus the situations which are of greatest interest to the petroleum industry are where charge builds up in or around flammable liquids and the possibility of discharge leading to incendiary sparking, and perhaps to a serious fire or explosion.

Countermeasures designed to prevent accumulation of electrostatic charges on a container being filled such as container grounding (ie, earthing) and bonding are routinely used. However, it has been recognized that grounding and bonding alone are insufficient to prevent an electrostatic buildup in low-conductivity organic liquids [8].

Organic liquids such as distillate fuels are inherently poor electric conductors. Static charge accumulates in these fluids because the electric charge moves very slowly through these liquids and can take a considerable time to reach a surface which is grounded. Until the charge has been dissipated, a high surface-voltage potential can be achieved, which can create an incendiary spark, resulting in an ignition or an explosion.

The risk of static discharge ignition is further compounded by the newly enacted legislation designed to improve the emission characteristics from combustion of fossil fuels.

To meet emission and fuel efficiency goals, researchers are investigating the use of nitrogen oxide traps, particulate traps, and direct injection technologies. Such traps and catalyst systems tend to be intolerant to sulfur. This coupled with the demonstrated adverse environmental consequences of burning sulfur-rich fuels has resulted in a global effort to reduce the sulfur content of fuels.

Ultra low sulfur fuels are becoming increasingly necessary to ensure compliance with emissions requirements over the full useful life of the latest technological generation of vehicles. Governments are also introducing further legislation for the reduction of particulate matter and fuel emissions.

10.3.1 Hydrotreating

The method most commonly used to reduce the sulfur content of fuels is hydrotreating [8]. Hydrotreating is a process by which hydrogen, under pressure, in the presence of a catalyst reacts with sulfur compounds in the fuel to form hydrogen sulfide gas and a hydrocarbon. However, hydrotreating to reduce sulfur content results not only in the removal of sulfur from the fuel but also in the removal of other polar compounds which normally increase the conductivity characteristics of the fuel.

Generally, a non-hydrotreated fuel has conductivity in the range of about 10 pS m^{-2} to about 30 pS m^{-2}, whereas the conductivity of a hydrotreated fuel (below the 15-ppm limit) is normally below 1 pS m^{-2}. A conductivity below 3 pS m^{-2} greatly increases the risk of catastrophic electrostatic ignition.

To correct the detrimental effects of hydrotreating, refineries and fuel handlers routinely use static dissipaters and conductivity improvers. When used properly, these additives minimize the risk of electrostatic ignition in hydrocarbon fuels and solvents.

10.3.2 Fuel Lubricity

Fuel lubricity is the ability of the fuel to prevent wear on contacting metal surfaces. Certain diesel engine designs rely on fuel as a lubricant for their internal moving components [8]. The problem of poor lubricity in these fuels is likely to be exacerbated by future engine system developments aimed at further decreasing emissions. This will result in an increase in the fuel oil lubricity requirement relative to the requirements for present engines. For example, the use of high-pressure unit injectors will likely increase the need for better fuel oil lubricity. Fuel lubricity requirements can be achieved by the use of lubricity additives.

As a consequence of the refinery processes used to reduce diesel sulfur and aromatics content, the majority of ultra low sulfur Diesel fuels marketed today will require treatment with additives to restore fuel lubricity and fuel electrical conductivity.

Many additive producers and additive users are combining lubricity and conductivity additives into multipurpose packages to address the low lubricity and low conductivity problems associated with ultra low sulfur fuels. Ostensibly, these combination packages not only deliver the required additives, but also should enable the additive user to eliminate the requirement of maintaining two separate additive addition pumps and storage containers. However, these combination packages fall short of the desired product [8].

The pump systems used by the fuel industry to deliver additives into fuels have great difficulty in accurately delivering these additives at such low treatment rates. Therefore fuel handlers using a single additive (conductivity) package are commonly required to dilute the additive (usually with hydrocarbon solvents or fuels) before injection of the conductivity additive into the fuel.

In the combined packages, the dilution of the conductivity additive is achieved by use of the lubricity additive as the diluent. The lubricity additive can be used as a diluent because the amount of lubricity additive required to treat a fuel is generally 50–100 times the amount of conductivity additive required to treat the same fuel [8].

The regular practice in the oil refining and fuel additives industries of combining two or more additives to provide a multipurpose package comes with certain precautions and requirements. It is, therefore, critical that the additives present in the fuel not only correct the specific fuel problems, or enhance the desired fuel attribute for which their addition is intended, but also do not have a detrimental effect on other fuel properties, or on the performance of other additives present in the fuel. This requirement is commonly referred to as *no harm*. The fuel and additives industries have developed a wide range of tests to evaluate the no-harm performance of additive packages and components [8].

10.3.3 Conductivity Improvers

Suitable static dissipaters and conductivity improver additives are summarized in Table 10.2.

Table 10.2 Conductivity improver additives

Compound	References
Aliphatic amines-fluorinated polyolefins	[9]
α-Olefin-sulfone copolymers	[10]
Polysulfone and quaternary ammonium salt amine/epichlorhydrin adduct dinonylnaphthylsulphonic acid	[11]
Copolymer of an alkyl vinyl monomer and a cationic vinyl monomer	[12]
α-Olefin-maleic anhydride copolymers	[13]
α-Olefin-acrylonitrile	[14, 15]
α-Olefin-acrylonitrile copolymers and polymeric polyamines	[16]
Copolymer of an alkyl vinyl monomer and a cationic vinyl monomer and polysulfone	[17]
Ethoxylated quaternary ammonium compound	[18]
Hydrocarbyl monoamine or hydrocarbyl-substituted polyalkyleneamine	[19]
Acrylic-type ester-acrylonitrile copolymers and polymeric polyamines	[20, 21]

Source: Burgazli CR, Henry Jr CP. Synergistic additive composition for petroleum fuels. US Patent 8 821 594, assigned to Innospec Fuel Specialities LLC (Newark, DE); 2014. http://www.freepatentsonline.com/8821594.html.

10.4 ENVIRONMENTAL CATALYSIS

Environmental catalysis has continuously grown in importance over the last 2 decades [22]. The development of innovative environmental catalysts is the crucial factor toward the objective of developing a new sustainable industrial chemistry.

Considerable expansion of the traditional area of environmental catalysis, mainly NO_x removal from stationary and mobile sources, and the conversion of volatile organic compounds has occurred. New areas include [22]:

- catalytic technologies for liquid or solid waste reduction or purification;
- use of catalysts in energy-efficient catalytic technologies and processes;
- reduction of the environmental impact in the use or disposal of catalysts;
- new eco-compatible refinery, chemical, or nonchemical catalytic processes;
- catalysis for greenhouse gas control;
- use of catalysts for user-friendly technologies and reduction of indoor pollution;
- catalytic processes for sustainable chemistry;
- reduction of the environmental impact of transport.

Therefore a significant change has occurred in the last decade in the areas of interest regarding environmental catalysts and in the way of approaching the research [22].

REFERENCES

[1] Engel DB, Goliaszewski AE, McDaniel CR. Separatory and emulsion breaking processes. US Patent 7 771 588, assigned to General Electric Company (Schenectady, NY); 2010. http://www.freepatentsonline.com/7771588.html.

[2] Lassila KR, Uhrin PA. Acetylenic diol ethylene oxide/propylene oxide adducts and processes for their manufacture. US Patent 6 313 182, assigned to Air Products and Chemicals, Inc. (Allentown, PA); 2001. http://www.freepatentsonline.com/6313182.html.

[3] Lassila KR, Uhrin PA, Schwartz J. Acetylenic diol ethylene oxide/propylene oxide adducts and processes for their manufacture. US Patent 6 864 395, assigned to Air Products and Chemicals, Inc. (Allentown, PA); 2005. http://www.freepatentsonline.com/6864395.html.

[4] Taylor RS, Stempler PS, Lemieux A, Funkhouser GP, Fyten GC, Cheng A, et al. Prevention of refinery plugging by residual oil gellant chemicals in crude-optimization of phosphonate ester oil gellants. J Can Petrol Tech 2006;45(05). http://dx.doi.org/10.2118/06-05-TN1.

[5] Delgado E, Keown B. Low volatile phosphorous gelling agent. US Patent 7 622 054, assigned to Ethox Chemicals LLC (Greenville, SC); 2009. http://www.freepatentsonline.com/7622054.html.

[6] Taylor RS, Funkhouser GP. Methods and compositions for treating subterranean formations with gelled hydrocarbon fluids. US Patent 6 511 944, assigned to Halliburton Energy Services, Inc. (Duncan, OK); 2003. http://www.freepatentsonline.com/6511944.html.
[7] Taylor RS, Funkhouser GP. Methods and compositions for treating subterranean formations with gelled hydrocarbon fluids. US Patent 6 544 934, assigned to Halliburton Energy Services, Inc. (Duncan, OK); 2003. http://www.freepatentsonline.com/6544934.html.
[8] Burgazli CR, Henry Jr CP. Synergistic additive composition for petroleum fuels. US Patent 8 821 594, assigned to Innospec Fuel Specialities LLC (Newark, DE); 2014. http://www.freepatentsonline.com/8821594.html.
[9] Bialy JJ, Siegart WR, Blackley WD, Chafetz H. Antistatic fuel composition. US Patent 3 652 238, assigned to Texaco Inc. (New York, NY); 1972. http://www.freepatentsonline.com/3652238.html.
[10] Johnson DD. Antistatic additive compositions. US Patent 3 811 848, assigned to E.I. Du Pont de Nemours and Company (Wilmington, DE); 1974. http://www.freepatentsonline.com/3811848.html.
[11] Henry Jr CP. Compositions of olefin-sulfur dioxide copolymers and polyamines as antistatic additives for hydrocarbon fuels. US Patent 3 917 466, assigned to E.I. Du Pont de Nemours & Co. (Wilmington, DE); 1975. http://www.freepatentsonline.com/3917466.html.
[12] Schield JA. Anti-static additives for hydrocarbons. US Patent 5 672 183, assigned to Petrolite Corporation (St. Louis, MO); 1997. http://www.freepatentsonline.com/5672183.html.
[13] Thompson NES. Antistatic agents for organic liquids. US Patent 4 416 668, assigned to Petrolite Corporation (St. Louis, MO); 1983. http://www.freepatentsonline.com/4416668.html.
[14] Naiman MI, Buriks RS. Olefin-acrylonitrile copolymers and uses thereof. US Patent 4 333 741, assigned to Petrolite Corporation (St. Louis, MO); 1982. http://www.freepatentsonline.com/4333741.html.
[15] Naiman MI, Buriks RS. Olefin-acrylonitrile copolymers and uses thereof. US Patent 4 388 452, assigned to Petrolite Corporation (St. Louis, MO); 1983. http://www.freepatentsonline.com/4388452.html.
[16] Naiman MI, Knepper JI, Thompson NES. Antistats containing acrylonitrile copolymers and polyamines. US Patent 4 259 087, assigned to Petrolite Corporation (St. Louis, MO); 1981. http://www.freepatentsonline.com/4259087.html.
[17] Schield JA. Anti-static additive compositions for hydrocarbon fuels. US Patent 6 391 070, assigned to Baker Hughes Incorporated (Houston, TX); 2002. http://www.freepatentsonline.com/6391070.html.
[18] Mor E. Electrostatic dissipative composition. US Patent 5 863 466; 1999. http://www.freepatentsonline.com/5863466.html.
[19] Wilkes MF, Duncan DA, Carney SP. Anti-static lubricity additive ultra-low sulfur diesel fuels. US Patent 6 793 695, assigned to The Lubrizol Corporation (Wickliffe, OH); 2004. http://www.freepatentsonline.com/6793695.html.
[20] Naiman MI. Antistats containing acrylonitrile copolymers and polyamines. US Patent 4 537 601, assigned to Petrolite Corporation (St. Louis, MO); 1985. http://www.freepatentsonline.com/4537601.html.
[21] Naiman MI. Antistats containing acrylonitrile copolymers and polyamines. US Patent 4 491 651, assigned to Petrolite Corporation (St. Louis, MO); 1985. http://www.freepatentsonline.com/4491651.html.
[22] Centi G, Ciambelli P, Perathoner S, Russo P. Environmental catalysis: trends and outlook. Catalysis Today 2002;75(1–4):3–15. http://www.sciencedirect.com/science/article/pii/S0920586102000378, http://dx.doi.org/10.1016/S0920-5861(02)00037-8.

CHAPTER 11
Processes

11.1 TYPICAL PROCESSES IN A REFINERY

There are monographs on processes that occur in refineries [1–5].

The processes in a refinery can be classified into physical methods and chemical methods. Distillation, solvent deasphaltation, solvent dewaxing, and solvent extraction are physical separation methods. Reforming, hydrotreating, cracking, isomerization, and alkylation are chemical conversion processes. In addition, thermal conversion processes include coking and visbreaking.

In a typical refinery, a number of different refinery feedstocks are processed, such as a number of different crude oils. The value of a feedstock will depend on the yield, composition, and properties of the distillate fractions it produces in a given refinery for subsequent refinery process streams and product blend components [6]. Typical processes that are used in a refinery are summarized in Table 11.1.

The refinery feedstocks are also usually blends of available feeds, and thus it is very difficult to predict the value of a feedstock in the overall refinery process, including detailed product quality and yield.

High-throughput methods that can be applied to the screening and the optimization of refinery processes have been developed [6].

11.2 DESALTING OF CRUDE OIL

In an oil refinery, the desalting of crude oil has been practiced for many years. The crude oil is usually contaminated from several sources, including [7]:

- brine contamination in the crude oil as a result of the brine associated with the oil in the ground;
- minerals, clay, silt, and sand from the formation around the oil well bore;
- metals, including calcium, zinc, silicon, nickel, sodium, and potassium;

Guide to the Practical Use of Chemicals in Refineries and Pipelines
http://dx.doi.org/10.1016/B978-0-12-805412-3.00011-8

Table 11.1 Typical refinery processes

Catalyzed processes	Uncatalyzed processes
Hydrotreating	Desalting
Isomerization	Distillation
Dehydrogenation	Membrane extraction
Aromatic saturation	Solvent extraction
Hydrocracking	Thermal cracking
Hydrogenation	Coking
Catalytic cracking	Coke calcining
Reforming	Bitumen blowing
Isodewaxing	Gasification
Sweetening	
Dealkylation	
Transalkylation	
Etherification	
Olefin dimerization	

Source: Butler G, Couves JW, Greenough P, Gudde NJ, Hodges MG. Process for evaluating a refinery feedstock. US Patent 8 546 146, assigned to BP Oil International Limited (Middlesex, GB); 2013. http://www.freepatentsonline.com/8546146.html.

- nitrogen-containing compounds, such as amines used to scrub H_2S from refinery gas streams in amine units or amines used as neutralizers in crude unit overhead systems, and also H_2S scavengers used in the oilfield;
- iron sulfides and iron oxides resulting from pipeline and vessel corrosion during production, transport, and storage;
- reactive phosphorus species that may result from gel compounds used in oil well stimulation.

Desalting is necessary before further processing to remove these compounds and other inorganic materials that would otherwise cause fouling and deposits in downstream heat exchanger equipment and/or form corrosive salts detrimental to crude oil processing equipment. Further, these phosphorus compounds and metals can act as poisons for the catalysts used in downstream refinery units. Effective crude oil desalting can help minimize the effects of these contaminants on the crude unit and downstream operations. Proper desalter operations provide the following benefits to the refiner [7]:

- reduced crude unit corrosion,
- reduced crude preheat system fouling,
- reduced potential for distillation column damage,

- reduced energy costs,
- reduced downstream process and product contamination.

11.2.1 Desalting Process

Desalting is the resolution of the natural emulsion of water that accompanies the crude oil by the creation of another emulsion in which about 5% relative wash water is dispersed into the oil with use of a mix valve. The emulsion mix is directed into a desalter vessel containing a parallel series of electrically charged plates. Under this arrangement, the oil and water emulsion is exposed to the applied electric field. An induced dipole is formed on each water droplet within the emulsion that causes electrostatic attraction and coalescence of the water droplets into larger and larger droplets. Eventually, the emulsion resolves into two separate phases—the oil phase (top layer) and the water phase (bottom layer). The streams of desalted crude oil and effluent water are separately discharged from the desalter.

The entire desalting process is a continuous flow procedure as opposed to a batch process. Normally, chemical additives are injected before the mix valve to help resolve the oil and water emulsion in addition to the use of electrostatic coalescence. These additives effectively allow small water droplets to more easily coalesce by lowering the oil-water interfacial tension.

Crude oil that contains a high amount of particulate solids can complicate the desalting process. The particulate solids would prefer to transfer to the water phase; however, much of the solids in a crude oil from a field exists in tight water-in-oil emulsions.

Thus oil-wetted solids in high concentration in the crude may form tight oil and water emulsions that are difficult to resolve. These tight emulsions are often referred to as *rag* and may exist as a layer between the separated oil and water phases. The rag layer inside the desalter vessel may grow to such an extent that some of it will be inadvertently discharged with the water phase. This is a problem for the wastewater treatment plant because the rag layer still contains a high percentage of unresolved emulsified oil [7].

Much of the solids encountered during crude oil desalting consists of particulates such as iron oxide, iron sulfide, sand, clay, and even phosphorus-containing compounds. Other metals that should be desirably removed include calcium, zinc, silicon, nickel, sodium, and potassium.

Some of the materials may be present in a soluble form, and some may require modification through a reaction such as hydrolysis or neutralization to become soluble. The metals may be present in inorganic or organic forms. In addition to complicating the desalter operation, phosphorus and other contaminants are of particular concern to further downstream processing. This includes the coking operation because iron and other metals remaining in the processed hydrocarbon yield a lower grade of coke. Removal of the metals from the crude oil early in the hydrocarbon processing stages is desired to eventually yield high-quality coke as well as to limit corrosion and fouling processing problems [7].

When crude oils or wash waters containing calcium and magnesium salts are processed at typical crude oil furnace temperatures, gaseous hydrochloric acid is evolved [8]. The so-formed hydrochloric acid may cause corrosion on the contact surfaces of processing equipment. To avoid the evolution of hydrochloric acid, it has become common practice to inject caustic into desalted crude oil so as to precipitate the calcium and magnesium cations contained in the oil as hydroxides while forming less hydrolyzable but still potentially corrosive sodium chloride. It is estimated that about 50% of domestic refiners inject caustic into desalted crude oil. Treatment with a sulfomethylated polyacrylamide may be helpful in these instances.

Several treatment approaches have been adopted to reduce total contaminant levels and these all center on the removal of contaminants at the desalter unit [7]. Normally, the desalter removes only water-soluble inorganic salts such as sodium or potassium chlorides. Some crude oils contain water-insoluble forms of phosphorus, which are soluble or dispersed as fine particulate matter in the oil but not in water.

In addition, some refineries have experienced the fouling of tower trays with phosphorus-containing deposits. The source of these phosphorus deposits may be the gel compounds used in oil well stimulation.

11.2.2 Membrane Separation

A comparison between two different wastewater treatments has been done to optimize the reuse of water and the recovery of heavy metal ions, such as Zn, Se, As, Cd, and Pb [9].

First, metal precipitation using sulfides was evaluated in terms of metal removal, amount of sludge produced, and water reuse. Then, a nanofiltration method was studied as an alternative separation process. This method was integrated in a precipitation process. Thus a water stream can be obtained that can be directly reused in the production process [9].

11.2.3 Electrostatic Desalting Dehydration Test Method

An electrostatic desalting dehydration apparatus (EDDA) was used as a laboratory test method. The method is as follows [7]:

1. Add 800, 600, or 400 mL of crude oil to be tested minus the amount of wash water (depending on the number of tubes the EDDA will hold) to a Waring blender.
2. Add the required percentage of wash water to the blender to bring the total volume up to 800, 600, or 400 mL.
3. Mix for 30 s. The speed can be reduced if ΔP on the mix valve is low.
4. Pour the mixture into the EDDA tubes to just below the 100-mL line.
5. Place the tubes in the EDDA heating block, which is at the desired test temperature (99°C).
6. Add the desired quantity of demulsifier, in parts per million, to each tube. With every test, a blank must be run for comparison purposes.
7. Place the screw-top electrode in the tubes and allow heat the samples for approximately 15 min.
8. Tighten the caps and shake each tube 100–200 times then place then back in the heating block to reheat them for 5 min.
9. Place the electrode cover over the tubes and lock it in place. Make sure that there is good contact between the cover and the electrode caps.
10. Set the time for 5 min and keep the tubes at 1500–3000 V, depending on the test requirements.
11. At the end of the 5 min, pull the tubes out and check for the percent water drop. Also check the quality of the interface and the quality of the water and record it.
12. Repeat steps 9, 10, and 11 until the desired total residence time is achieved.
13. Determine the best candidates and run a dehydration test on those samples.
    - Fill the desired number of 12.5-mL centrifuge tubes to the 50% mark with xylene.
    - Use a glass syringe to pull 5.8 mL of dehydrated crude sample from the desired level in the tube and mix it in with the xylene in the centrifuge tubes.
    - Centrifuge the tubes at 2000 rpm for 4 min.
    - Record the quantities of water, emulsion, and solids that are present at the bottom of the tube.

11.2.4 Membranes

For many years it has been hoped that gas separation through a membrane will be an environmentally benign and simple process in a refinery. Namely, distillation, extraction, absorption, and adsorption processes are generally energy-consuming processes. A molecular sieve DDR-type zeolite membrane with an aperture of 0.36 nm × 0.44 nm was formed on a porous alumina substrate by a hydrothermal process [10]. The permeance through the membrane in the single gas feed of helium, hydrogen, carbon dioxide, oxygen, nitrogen, methane, *n*-butane, isobutane, and sulfur hexafluoride was measured at 301 and 373 K up to 0.5 MPa.

The permeance decreased by more than three orders of magnitude between 0.35- and 0.40-nm kinetic diameter of permeated gas at both 301 and 373 K [10]. The separation factors for CO_2 to CH_4 in 50% CO_2 and 50% CH_4 mixed gas feed were 220 and 100 at 301 and 373 K, respectively, at 0.5 MPa total gas feed pressure. The permeation properties were slightly affected by the water addition at 301 K. It was concluded that the DDR-type membrane had only a few defects and worked well as a molecular-sieving membrane.

11.2.5 Chemicals

Acetic acid has sometimes been used to remove metals, but it has a high oil solubility and tends to stay with the hydrocarbon coming from the desalter [11]. Further, the acidity of acetic acid can cause corrosion problems in the crude unit.

Nonyl phenol resins have been used as desalting additives, but these materials are suspected of being hormonal mimics and are ineffective by themselves in removing metals, such as calcium or iron [7].

Thioglycolic acid and the ethers of glycolic acid may have the added benefits of a higher boiling point and possibly increased water solubility [11]. A higher boiling point means the additive will not distill into the distillate fractions in the crude unit and cause corrosion or product quality concerns. The higher water solubility also favors removal of the additive from the crude oil in the desalter and reduces the amount that may reach the downstream processing units. The addition of glycolic acid (ie, hydroxyacetic acid) and other water-soluble hydroxyacids to a crude oil can significantly reduce the amount of calcium and other metals and amines as well as reactive phosphorus species in the hydrocarbon when it is run through a desalter in a refinery [7].

The addition of glycolic acid at levels of up to a 5:1 ratio with calcium results in much lower metal and amine contents of the desalted oil. Also, the levels of metals other than calcium, such as iron, zinc, silicon, nickel, sodium, and potassium, are reduced.

For testing, a crude oil with a high calcium content was used. As a first additive, 70% glycolic acid in aqueous solution was used. Further, an additive consisting of a blend of glycolic acid, phosphoric acid (as a pH adjuster), a pyridine quaternary ammonium compound (as a corrosion inhibitor), a dinonyl phenol/ethylene oxide oxyalkylate (cosolvent), isopropyl alcohol, and water was used. The results of testing by the EDDA method described above are shown in Table 11.2.

A lot of related additive formulations have been tested and the results have been given in the literature [7].

In addition, a cross-linked polypropylene glycol has been used [12]. Suitable cross-linking agents include, bisphenol A, epoxy resins, anhydrides, and toluene diisocyanate.

Table 11.2 EDDA testing results

Metal	**Content (ppm)**			
	Crude oil	**Top phase**	**Interface**	**Water phase**
Additive: 70% glycolic acid in aqueous solution				
Calcium	370	30	31	1700
Iron	60	14	15	113
Zinc	35	6	4	163
Silicon	37	<2	<2	6
Nickel	8	9	9	<2
Sodium	97	9	10	416
Potassium	789	31	32	4030
Additive: glycolic acid, phosphoric acid, pyridine quaternary ammonium dinonyl phenol/ethylene oxide oxyalkylate, isopropyl alcohol, water				
Calcium	370	76	76	1210
Iron	60	26	27	8
Zinc	35	17	16	34
Silicon	37	<2	2	7
Nickel	8	9	10	<2
Sodium	97	13	12	404
Potassium	789	34	32	3900

Source: Kremer LN, Nguyen TM, Weers JJ. Additives to enhance phosphorus compound removal in refinery desalting processes. US Patent 7 799 213, assigned to Baker Hughes Incorporated (Houston, TX); 2010. http://www.freepatentsonline.com/7799213.html.

Phosphorus-containing salts that form fouling deposits on the trays of refineries may be made by the mixing of alkyl phosphate esters with aluminum-containing compounds. The reaction results in a three-dimensional structure that is a basis of the gels used in oil well stimulation. The addition of glycolic acid will help to promote the hydrolysis of phosphate ester gelling compounds because esters are known to be hydrolyzed by acid catalysts [7].

11.3 FOULING MITIGATION

In petroleum refineries additional energy costs may arise from fouling and the resulting inefficiencies caused by the fouling processes [13]. In particular, thermal processing of crude oils, blends, and fractions in heat transfer equipment, such as heat exchangers, is hampered by the deposition of insoluble asphaltenes and other contaminants that are inherent in most crude oils. Further, the asphaltenes and other organics are known to thermally degrade to coke when exposed to high heater tube surface temperatures.

In addition, because crude oils are typically blended at the refinery, the mixing of certain types of crude oils can create a foulant [14]. The asphaltenic material that is precipitated by the blending of incompatible crude oils will often lead to a predominantly organic type of fouling, which with prolonged heating will form a carbonaceous or coke-like foulant deposit. Crude oils often also contain acidic components that directly corrode the heat exchanger materials as well. Naphthenic acids will remove metal from the surface and sulfidic components will cause sulfidic corrosion, results in the formation of iron sulfide. This sulfidic scale that is formed is referred to as sulfide-induced fouling.

Fouling in heat exchangers receiving petroleum-type process streams can result from a number of mechanisms, including [13]:

- chemical reactions,
- corrosion,
- deposit of existing insoluble impurities in the stream, and
- deposit of materials rendered insoluble by the temperature difference between the process stream and the heat exchanger wall.

For example, naturally occurring asphaltenes may precipitate from the crude oil process stream, thermally degrade to form a coke, and adhere to the hot surfaces. Further, the high ΔT inherent in a heat transfer operation results in high surface or skin temperatures when the process stream is introduced to the heater tube surfaces, which contributes to the precipitation of insoluble particulates.

Another common cause of fouling is attributable to the presence of salts, particulates, and impurities (eg, inorganic contaminants) found in the crude oil stream. Iron oxide, iron sulfide, calcium carbonate, silica, sodium chloride, and calcium chloride have all been found to attach directly to the surface of a fouled heater rod and throughout the coke deposit. These solids may promote and enable additional fouling of the crude oils.

The buildup of insoluble deposits in heat transfer equipment creates an unwanted insulating effect and reduces the heat transfer efficiency. Fouling also reduces the cross-sectional area of the process equipment, which decreases the flow rates and the desired pressure differentials, resulting in a reduced operation performance. To overcome these disadvantages, heat transfer equipment must be regularly taken off-line and cleaned mechanically or chemically, thus resulting in a loss of production time.

Thus it is desirable to reduce the precipitation and the adherence of particulates and asphaltenes from the heated surface to prevent fouling, and this should be done before the asphaltenes are thermally degraded or coked. This precaution will improve the performance of the heat transfer equipment, decrease or eliminate scheduled outages for fouling mitigation efforts, and reduce the energy costs associated with the processing methods [13].

Metal sulfonate additives have been developed for fouling mitigation in petroleum refinery processes [13]. For example, calcium sulfonates can be obtained from Infineum (Abingdon, United Kingdom, and Linden, NJ, United States). A preferred additive is Infineum® C9350. It is a long-chain alkylbenzene sulfonate. It is a nonflammable, nonexplosive, viscous brown liquid with a faint petroleum odor and low water solubility. Further, Infineum® C9350 can be used as a detergent additive in crankcase motor oils and as an additive to oils and as a lubricant in the cutting of metals [13].

The synthesis of additives of this type has been described in the literature [15–17]. Exemplary synthesis techniques have been described [18].

In compositions for antifouling agents, solubilizers are also added, such as nitrogen-containing phosphorus-free carboxylic solubilizers [19].

As dispersants, boric acid and organic derivatives of boric acid may be added. It is believed that the metal cation moiety and the borate moiety synergistically combine to increase the effectiveness of the resulting product [13].

Compositions for use as antifouling agents have been described [13]. Several commercially available metal sulfonates have been blended with organic borates at elevated temperatures to form a series of products with high boron content. The efficiency of the formulations has been tested in an Alcor Hot Liquid Process Simulator.

11.3.1 Corrosion-Resistant Materials for Reduced Fouling

Heat transfer components are used in refinery and petrochemical processing applications at various locations within the facilities to adjust the temperature of the processed fluid. The heat transfer components are heat exchangers, furnaces, and furnace tubes located in refining facilities.

A typical tube-in-shell heat exchanger includes a plurality of tubes which the oil may flow through and around. A hot fluid and a cold fluid enter separate chambers or tubes of the heat exchanger unit. The hot fluid transfers its heat to the cold fluid. The heat exchanger is designed to efficiently transfer heat from one fluid to another. The hot and cold fluids are never combined. Heat transfer occurs through the tube wall that separates the hot and cold liquids.

A variety of other heat exchanger designs, such as spiral heat exchangers, tube-in-tube heat exchangers, and plate-and-frame heat exchangers, operate essentially on the same principles.

During normal use with contact between the oil and the heat exchanger, corrosion and the buildup of deposits may occur. This buildup of deposits is often called fouling. Fouling adversely impacts the optimal control of the heat exchanger.

The loss in heat transfer efficiency results in higher fuel consumption at the furnace and reduced throughput. The buildup of foulants in fluid transfer components results in reduced throughput, higher loads on pumping devices, and plugging of downstream equipment. Also, the surfaces of the tubes of a heat exchanger are subject to corrosion.

Improved heat transfer components which exhibit improved corrosion resistance and improved resistance to fouling have been developed [14].

Heat exchanger tubes may be fabricated from an aluminum- or aluminum alloy-coated carbon steel or a steel composition that is resistant to sulfidation or sulfidic corrosion and fouling. The use of aluminum- or aluminum alloy-coated carbon steel or a steel composition that is resistant to sulfidation and fouling significantly reduces fouling and corrosion.

The steel consists of several layers. The primary layer is an iron steel with low amounts of chromium. The secondary layer is a chromium-enriched steel with more than 10% chromium.

The protective layer includes a magnetite, an iron-chromium spinel, and a chromium oxide. This layer may also contain a mixed oxide sulfide thiospinel [14]. The protective layer is formed in situ within the heat exchanger. It is preferably formed by exposure of the chromium-enriched layer to a process stream at high temperatures.

11.3.2 Analysis of Foulants

Most refinery fouling is the result of some common causes that can be detected and verified by determination of the chemical composition by a standardized analytical testing. A detailed procedure has been presented in the literature [20]. Fouling causes can be divided into five general classes, which are summarized in Table 11.3.

Fouling often results from a combination of a series of the causes given in Table 11.3. Hence characteristics have been determined which can identify each of these common causes of fouling on the basis of the chemical composition of the deposit. As a result, a sequence of analytical tests for the foulant recovered from the fouled unit were designed to diagnose the causes of fouling [20].

It is extremely important that accurate records are made of the location and approximate quantity of foulant as well as the method conditions during the fouling and before the fouling was noticed. Pictures of the foulant on the unit are also very helpful. Important method conditions needed for the root-cause determination include temperature, pressure, flow rate (residence time), source of the method fluids, and details on the geometry of the unit.

For example, if a foulant is obtained from a heat exchanger, it is important to know if the sample came from within the tubes, from the tubesheet face, or from an unheated surface of the channel box. This knowledge, which include knowledge of surface temperatures, metallurgy, and velocity factors, can provide valuable clues to the foulant source compared with the simple statement that the foulant was retrieved from a particular exchanger.

Care should be taken to ensure that the foulant sample is representative and not contaminated by collection or flushing procedures. If it is also clear that there are different types of foulants, they should each be sampled. Ideally, about 100 g of foulant should be collected for each sample. If this is not possible, as much sample as possible should be collected. The sample should be stored in a closed container until the analysis is performed.

Table 11.3 Classes of fouling

Main class	Examples
Scaling: precipitation of salts	Sea salts from crude oil containing dispersed water Deposition from cooling system water
Corrosion fouling: the reaction of the heat transfer surface to form corrosion products	Iron sulfide formation by reactive sulfur or hydrogen sulfide Iron naphthenate formation by naphthenic acids in the oil and possible subsequent reactions with hydrogen sulfide to form iron sulfide
Precipitation fouling: formation of an insoluble phase by cooling, by mixing two or more streams, or by breaking an emulsion within the fouled unit	Wax deposition by cooling below the freezing temperature of high molecular weight paraffins Asphaltene deposition Mixing incompatible oils in the wrong proportions By breaking an emulsion
Chemical reaction fouling: formation of an insoluble phase by chemical reaction between components in the fluid	Thermal coking (asphaltene insolubility during thermal cracking) Polymerization of conjugated olefins Aromatic growth and oxidation Ammonium chloride and bisulfide formations
Particulate fouling: dispersed particles formed upstream and carried over to and deposited within the fouled unit	Coke Asphaltenes Iron sulfide Catalyst fines

Source: Brons G, Brown LD, Joshi H, Kennedy RJ, Bruno T, Rudy TM. Method for refinery foulant deposit characterization. US Patent 7 708 864, assigned to ExxonMobil Research & Engineering Company (Annandale, NJ); 2010. http://www.freepatentsonline.com/7708864.html.

11.3.3 Analysis of Solid Foulant Samples

The first step is to allow 10–25 g of the sample to soak for at least 4 h in toluene or another solvent with similar solvent capability to dissolve the trapped feed from the foulant [20]. After filtration, the remaining solids are washed with additional solvent until a clear filtrate is obtained. The insoluble solids are then dried in a vacuum oven at 70°C for at least 1–2 h. If the entire foulant dissolves in the solvent, the procedure should be repeated on another sample of the foulant, with toluene replaced by *n*-heptane to obtain insoluble asphaltenes. This washing procedure is important so that subsequent results can be ascribed to the foulant rather than a mixture of the foulant and the trapped feed.

It is important that a homogeneous sample be obtained before the thermogravimetric analysis (TGA) and elemental/metals analysis. One can achieve this by grinding the foulant sample to −60 mesh and mixing it thoroughly before the analytical tests are performed. Often this is best done after the toluene extraction. However, if the foulant is a hard, granulated solid (not paste-like), this may be done before washing with toluene.

Before the above-mentioned grinding and homogenization of the entire sample, a toluene-insoluble, unground portion of the foulant is subjected to scanning electron microscopy and energy-dispersive X-ray analysis to first qualitatively identify the elements present and their association with each other. Scanning electron microscopy is an essential step in this method. The information obtained is used to identify which elements are needed for subsequent quantification of the toluene-insoluble, ground and homogenized sample to determine the amount of different materials that may be present in the foulant deposit [20].

The first part of the analysis is to determine how much of the insoluble foulant is due to inorganics (eg, metals, salts, and corrosion products) and how much is due to organic compounds. Also, the carbon, hydrogen, and nitrogen contents in samples of coal and carbon can be determined according to standard test methods [21].

11.3.4 Inorganic Fouling

The initial step after scanning electron microscopy and energy-dispersive X-ray analysis is to determine how much of the insoluble sample is due to inorganics [20]. This can be achieved with an elemental analysis for carbon, hydrogen, sulfur, nitrogen, chlorine (halogens), and metal contents and an ash test, which is part of the TGA. Analysis for oxygen and other elements may also be done if the other elements add up to significantly less than 100% or if there is reason to suspect that they are present.

If the sample contains 10–15% or greater inorganics and it is confirmed that it was not part of the metal surface of the unit removed with the foulant, inorganic fouling should be addressed first. Although there can be more than one cause, the deposition of inorganics can also promote the deposition of organics by allowing increased surface areas for asphaltenes (eg, for laying down and thermal degradation to coke). Elemental analysis will suggest if the primary cause is due to iron sulfide (common), sea salts, ammonium chloride, aluminum silicate, or compounds.

11.3.5 Organic Fouling

If the toluene-insoluble portion of the foulant sample is less than 10–15% inorganic, the bulk of the foulant may be due to coke or high molecular weight waxes [20]. The cause of the organic fouling should be determined.

The composition of the inorganics may be used to trace the origin of the organic fouling precursors. For example, vanadium and nickel are naturally connected to organics as porphyrins and are soluble in crude oils and resids. Thus these metals can indicate entrained resid in a gas oil stream, for example. Nevertheless, the next step is to determine by optical microscopy, with use of cross-polarized light, whether the foulant is isotropic (unordered) or anisotropic (ordered) coke. One determines whether or not the coke is ordered or unordered by the patterns observed with the microscope. For example, anisotropic (ordered) coke exists as distinct patterns of various shapes or types, while isotropic (unordered) coke has no distinguishable formation patterns.

Anisotropic coke derived from carbonaceous mesophase shows that it was formed by thermal cracking followed by phase separation of insoluble asphaltenes, such as the coking of resid. If the coke is isotropic, a possible cause is the initial polymerization of conjugated olefins. The presence of significant volatiles below 370°C in TGA may be an indication that the cause is polymerization. The other indication is that polymerization usually occurs in condensed liquid products of thermal cracking (such as from cokers) after being held at 232–324°C. Asphaltenes can also become insoluble by the blending of oils containing asphaltenes with oils high in saturates or the cooling of a thermally converted oil, such as a visbreaker tar. TGA volatiles between 500 and 800°C are typically observed when coke exists in the deposit. During combustion with oxygen at 800°C, additional coke is volatilized from the sample, leaving only ash which is due to metals in the deposit that now exist in an oxide state.

Finally, the hydrogen-to-carbon atomic ratio of the coke can be roughly correlated with the temperature of deposition This can help to determine if the coke was formed at a higher temperature upstream of the fouled unit, if the temperature was higher than measured, if the coke had been present for significant lengths of time, or if a catalytic mechanism was involved. It is known from the literature that asphaltenes have hydrogen-to-carbon atomic ratios in the 1.0–1.2 range, coke materials have lower hydrogen-to-carbon ratios, while the hydrogen-to-carbon atomic ratios of waxes can be as high as 2.0. It is also important to correlate hydrogen-to-carbon ratios with findings from microscopy. The hydrogen-to-carbon ratio is an average value of the homogenized sample, which may contain different types of coke.

If the foulant is soluble in toluene but insoluble in *n*-heptane it is an asphaltene by definition. Thus it is formed by one of the two mechanisms discussed above but was not heated enough for it to degrade to coke (coke is defined as toluene insoluble). While it is usually clear from the method conditions which mechanism applies, a general indication that the asphaltenes were thermally cracked is that they have a hydrogen-to-carbon atomic ratio less than 1 [20].

11.4 POLAR MOLECULE CONTAMINANTS

Many polar organic compounds such as sulfur-containing molecules, naphthenic acid, asphaltene, porphyrin, and nitrogen-containing molecules cause problems in the processing in refineries [22]. These polar molecules may cause corrosion, fouling, catalyst poisoning, and emissions. Therefore the safe and cost-effective removal of polar molecules from the hydrocarbon and chemical streams increases energy savings and the profitability of the process.

One can remove polar molecules from a fluid by directing the flow of the liquid through a fixed bed of particles which are capable of adsorbing the polar molecules. However, a fixed bed process generally precludes the use of very small adsorbent particles that are smaller than 0.5–1.0 mm because of the excessive pressure drop that will result in a commercial process when such particles are used. This high pressure drop becomes an even bigger concern if the fixed bed fouls and plugs up.

In addition, a fixed bed process requires the periodic regeneration of the fixed bed. Because of the poor thermal conductivity and large size of commercial fixed beds, the regeneration may require the heating of the bed for several hours to achieve a temperature high enough to regenerate the bed. The time required for regeneration of fixed beds thus results in an industrial operation that is neither practical nor economical [22].

Polar molecule contaminants can be removed from a refinery stream by magnetic nanoparticle compounds with a high surface area [22]. The magnetic nanoparticle may be formed from any material that can be attracted to a magnetic field (eg, iron, nickel, cobalt, or magnetite). The refinery stream is a liquid hydrocarbon stream in a hydrocarbon refining process. The separation of the nanoparticle onto which the polar molecule complex is adsorbed can be accomplished by the application of a magnetic field to the nanoparticle complex, thus separating the complex from the hydrocarbon liquid fluid.

Eventually, a step involving regeneration of the nanoparticle complex removes the polar molecule contaminant adsorbed onto the nanoparticle compound. The regeneration can be achieved by heating of the complex at 250°C or by immersion of the complex in water.

The concept has been demonstrated by the removal of asphaltene from toluene, the removal of porphyrin from toluene, and the removal of naphthenic acid [22]. For example, in the case of asphaltene removal, the initial concentration of asphaltene of 770 ppm could be reduced by 87.5% after the first treatment with magnetic nanoparticles, and was reduced by about 100% after a second treatment.

11.5 SOAPS FROM CHEMICAL PULPING

In a typical chemical pulp mill, such as a Kraft pulp mill, white liquor containing active cooking chemicals, NaOH and Na_2S, is used for cooking wood chips. The residue, which is called *black liquor*, is concentrated by evaporation and burned in a recovery boiler furnace to yield an inorganic smelt of Na_2CO_3 and Na_2S. The smelt is then dissolved to form a green liquor, which is reacted with $Ca(OH)_2$ to convert Na_2CO_3 to NaOH and to regenerate the original white liquor.

Instead of regeneration, as described above, the black liquor can be used as the base material for a refinery feedstock. A process for preparing a refinery feedstock from black liquor soap created from chemical pulping has been presented [23].

An excess of methanol is added to the black liquor soap. Acid is then added to the mixture to decrease the pH of the mixture to approximately 2 to convert carboxylate salts of fatty and resin acids to the free fatty and resin acids. The free fatty and resin acids react with the alcohol, with the aid of the acid catalyst, resulting in ester products. The resulting feedstock can be distilled or refined to yield sterols and related alcohols, biodiesel, and other fuels.

The refinery feedstock produced from this process will be stabler and more consistent than traditional plant-based refinery feedstock, which is dependent on growing seasons. This process also makes use of existing pulp mill facilities and therefore is considerably less capital intensive than known greenfield biodiesel plants requiring the installation of steam generating and acid facilities. In particular, the advantages of the above-described process are [23]:

1. Production of a refinery feedstock from pulp mill by-product, thereby providing a refinery feedstock with ready, year-round supply.
2. Reduction of the yield lost to esterification of resin and fatty acids with the sterols and related alcohols, by the conventional tall oil process.
3. Reduction of greenhouse gas emissions when the resulting feedstock is further refined and used as a fuel in place of fossil fuels.
4. Black liquor soap is mixed with methanol, thereby making it easier to pump and mix.
5. The yield of refinery feedstock is improved, with respect to the production of tall oil, because the side reaction of the fatty and resin acids with sterols and related alcohols in the black liquor soap is greatly reduced by addition of methanol in excess. The yield of sterols is improved as well.
6. The production of the refinery feedstock starts with a very low cost raw material (black liquor soap). Black liquor soap typically has no value to a mill. It is a waste product and is a disposal problem. Other processes which create biodiesel from tall oil have a much higher cost because of the value already added to the tall oil due to its previous processing.

11.6 CATALYTIC REFORMING

11.6.1 Upgrading Light Hydrocarbons

Catalytic reforming is one of the basic petroleum refining processes for upgrading light hydrocarbon feedstocks [24]. Products from catalytic reforming include high-octane gasoline useful as automobile fuel, aromatics, and hydrogen. The typical reactions in catalytic reforming are dehydrocyclization, isomerization, and dehydrogenation. Dealkylation and hydrocracking reactions are undesirable owing to the low value of the resulting light hydrocarbon products.

Catalysts commonly include platinum or palladium, or a group VIII metal plus a second catalytic metal, which acts as a promoter. Examples of promoter metals include rhenium, tin, tungsten, germanium, cobalt, nickel, rhodium, ruthenium, and iridium.

The catalytic metal may be dispersed on a support such as alumina, silica, or silica-alumina. Typically, a halogen such as chlorine is incorporated on the support to add an acid functionality.

Other reforming catalysts include molecular sieves, such as borosilicates and silicoaluminophosphates, layered crystalline clay-type phyllosilicates, and amorphous clays [24].

A multistage reforming process comprises the passage of a refinery stream through at least two reforming stages in series [24]. In general, each reforming stage is characterized by one or more reforming reactor vessels, each containing a catalyst and maintained at reforming reaction conditions. The product from each stage before the final stage is passed, at least in part, to the succeeding stage in the multistage process. Preferably the final stage is run at a lower pressure than the penultimate stage [24].

11.6.2 Reducing Benzene Content

A catalytic reforming process has been described that produces gasoline with a reduced benzene content. This process includes the steps of [25]:

1. reforming a reformer feedstock that includes a naphtha stream to produce a gasoline reformate product stream,
2. splitting the gasoline reformate product stream into one or more relatively benzene-rich fractions and one or more relatively benzene-lean fractions,
3. hydrogenating the one or more relatively benzene-rich fractions to produce a cyclohexane-rich effluent,
4. recycling at least a portion of the cyclohexane-rich effluent to constitute a portion of the reformer feedstock.

11.6.3 Continuous Catalytic Regeneration

When specifically selected coke precursor compounds from the front end of oil distillate fractions are used that contain C_{11}-C_{14} hydrocarbons, these may be used as additives for the processing of naphtha in a catalytic reformer [26]. The C_{11}-C_{14} hydrocarbons additives enhance the coke in continuous catalytic regeneration reformers to levels higher than those which are usually produced in low coke naphtha reforming operations.

Owing to the increased ethanol blending in gasoline and low reformate octane severity operations, reformers do not produce the necessary amount of coke to permit a steady state of the operations to be sustained [26].

11.6.4 Coupling With Water Purification

A catalytic steam-hydrocarbon reforming process can be coupled with a thermal water purification process [27]. A low-cost energy source for thermal water purification is a catalytic steam-hydrocarbon reforming process.

11.7 METHANOL REFINERY

A process for converting gas resources in remote locations to fuels and petrochemicals has been described [28]. Although the operating cost for producing fuels and petrochemicals from remote natural gas is significantly lower than that for producing the same products from oil, high capital costs discourage such a conversion of natural gas. Reducing such capital costs would provide an incentive for utilization of natural gas as a relatively inexpensive source of fuels and petrochemicals.

Natural gas is often coproduced with oil in remote offsite locations where the reinjection of the gas is either expensive or not feasible. A desirable option for treating such gas is its conversion to methanol, by means of the *Fischer-Tropsch* process, which is simpler than gas liquefaction.

Methanol can be produced at reasonable cost in plants ranging from 500 to 50,000 $t\,day^{-1}$. While larger plants have a large cost advantage, small plants are nevertheless viable if the gas cost is sufficiently low (eg, the reinjection of natural gas is too costly or impossible). Accordingly, it has been suggested as economically feasible to place methanol plants on barges for offshore producing locations [28].

Steam methane reforming is the catalytic reaction of natural gas with steam to produce a synthesis gas. The steam methane reformation reaction is endothermic; therefore external heat is required. The natural gas and steam are typically fed into alloy tubes that contain a nickel-based catalyst for the reforming reaction. The catalyst tubes are placed inside a refractory lined structure. A portion of the natural gas is used as fuel to provide the heat required for the reaction [28]. Then the production of methanol occurs by oxygenated catalytic reforming.

In the next step, methanol is shipped from the local synthesis plant to the methanol conversion refinery by any suitable means, such as dedicated large tankers, supertankers, or pipelines. The cost to ship the methanol is expected to be similar to the cost of shipping crude oil. In the methanol conversion refinery the conversion of methanol to gasoline occurs.

The conversion into olefinic compounds and aromatics can be achieved by zeolite-based catalysts [29, 30]. The addition of various promoters, including aromatic compounds, such as toluene, can accelerate the conversion of methanol to hydrocarbons over zeolites [31]. Preferably, olefin production from methanol is done with a silicon metal aluminophosphate molecular sieve catalyst [32, 33].

11.8 HYBRID REFINERY

Many methods have been suggested for utilizing biofuel for the production of energy in order to compensate for at least a portion of the fossil fuel currently used in such energy production, and thereby also decrease carbon dioxide emissions in the overall energy production cycle [34].

Biofeedstocks are generally considered to be low-energy fuels, and cannot easily be utilized for the production of energy. The low energy content of biomass renders biomass generally inadequate for high-efficiency production of energy, such as high-temperature, high-pressure steam, or electricity.

The nonuniformity in the composition of biomass causes differences in its quality and other similar hard-to-control variations. This may cause problems in an energy production cycle that relies heavily on such a biofuel.

Research has been directed to processes for introducing biomass into a conventional refinery. Such a process may comprise the steps of [34]:

1. co-feeding biomass and refinery residual material as a hybrid feedstock to a gasifier,
2. gasifying the hybrid feedstock in the gasifier to form synthesis gas,
3. processing the synthesis gas to form a derived product, usable in a conventional refinery.

Biomass may include agricultural feedstocks, forestry-based feedstocks, and municipal solid waste [34]. Municipal solid waste may contain waste plastics, used tires, paper, scrap wood, food-processing waste, sewage, sludge, or green waste.

Conventional biomass contains oils rich in triglycerides, which are widely used in industry, referred to as biomass oils, including vegetable oils and animal fat lipids which provide fatty acids by various reaction mechanisms [35]. Fatty acids have an even number of carbon atoms, both saturated and unsaturated. In oils, fatty acid glycerides with unsaturated carbon chains predominate. In contrast, in fats, saturated fatty acid glycerides predominate.

Some biomass oils can be used directly as fuels or processed into biodiesel of fatty acid esters. Also, they may be coprocessed in petroleum refining units to produce fuels with a contribution of hydrocarbons generated from renewable sources [35]. The coprocessing of biomass oils in petroleum refining is focused on the conversion of triglycerides into hydrocarbons, making it possible to advantageously use an established refinery scheme with minor modifications and with a potential benefit in terms of the cost of production and distribution of bioproducts on a small scale [35].

A major factor in conventional petroleum refinery hydrotreating processing units which makes the coprocessing of biomass oil difficult is the high exothermal hydroconversion reaction of the triglyceride and the generation of H_2O, CO, and CO_2 gases due to decarboxylation and the decarbonization reactions of the fatty acids [35]. An increase in the concentration of the CH_4, CO, and CO_2 gases in the reaction medium decreases the activity of the catalysts in conventional refinery streams. The major reason for the catalytic activity decrease in the reaction zone is due to a reduction in the H_2 partial pressure which occurs when there are increases in CH_4 concentrations in the gas recycle. These problems can be diminished to some extent by use of a special catalytic system [35]. This system consists of a first monometallic catalytic bed containing only one group VIB metal (Mo) oxide for the hydroconversion of the triglycerides into *n*-paraffins and a second bimetallic catalytic bed (NiMo) for hydrotreating the refinery stream. In this way, the triglyceride hydroconversion reactions are selectively controlled, minimizing the generation of undesirable gases mentioned above and increasing the conversion to the corresponding *n*-paraffins.

11.8.1 Pyrolysis of Biomass

Pyrolysis is the chemical decomposition of a condensed substance by heating. Pyrolysis is a special case of thermolysis and it is mostly used for organic materials. It occurs spontaneously at high temperatures (eg, above 300°C for wood). Pyrolysis generally does not involve reactions with oxygen but can occur in the presence of oxygen. Extreme pyrolysis, which leaves only carbon as the residue, is called carbonization and is also related to the chemical process of charring.

Since pyrolysis is endothermic, heat transfer considerations dominate the design of pyrolysis reactors. Direct heat transfer with a hot gas is possible in a pyrolysis reactor, but it is difficult to provide enough heat with reasonable gas flow rates because of the low specific heat of a relatively thin gas. Indirect heat transfer with exchange surfaces is also possible, but it is difficult to achieve good heat transfer on both sides of the heat exchange surface, and these methods are not easily scaled up.

Direct heat transfer with a circulating hot solid is possible. This is an effective method, but the technology for moving, recovering, and reheating the solid can be complex.

Rotating cones have also been used to introduce and mix hot solids with the comminuted biomass. Preheated hot sand and biomass particles are introduced into a rotating cone, and the mixture of sand and biomass is transported across the cone surface by centrifugal force due to the rotation.

Also, fluidized beds have been used, whereby biomass particles are introduced into a bed of hot sand fluidized by a gas, which is usually a recirculated product gas. High heat transfer rates from fluidized sand result in rapid heating of biomass particles. Circulating fluidized beds are yet another possibility. Biomass particles are introduced into a circulating fluidized bed of hot sand. Gas, sand, and biomass particles move together, with the transport gas usually being a recirculated product gas. High heat transfer rates from sand ensure rapid heating of biomass particles, and ablation is stronger than with regular fluidized beds. Although the above-mentioned processes may be scaled up, the process is rather complex.

Methods for pyrolyzing biomass with high conversion rates at lower temperatures that can be scaled up and are cost-effective are desired. A method which combines the processing of the biomass with a refinery feedstock before pyrolysis meets these needs. Such a method of preparing a hydrocarbon from biomass by pyrolysis has been described. The method consists of the steps of [36]:

1. mixing biomass particles together with a hot refinery feedstock that is at less than 500°C and serves as a heat source for pyrolyzing the biomass particles,
2. recovering a hydrocarbon oil pyrolysis product that is about 60% bio-oil, wherein the hot refinery feedstock serves as the sole heat source for the method.

The apparatus is shown in Fig. 11.1.

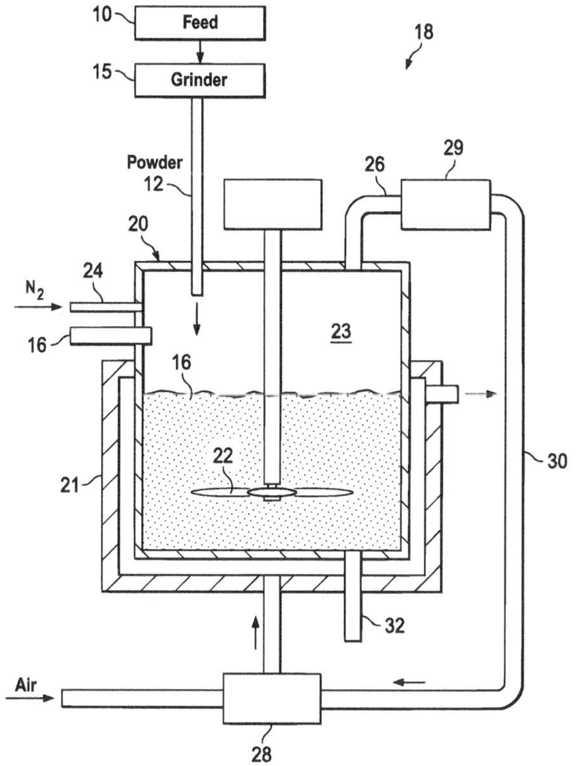

Figure 11.1 Biomass addition apparatus. *(Source: Jess A, Kaufmann D, Daugaard DE. Biomass pyrolysis in refinery feedstock. US Patent 8 287 723, assigned to Phillips 66 Company (Houston, TX); 2012. http://www.freepatentsonline.com/8287723.html).*

There, the biomass 10 is ground to a powder 12 in a suitable grinding apparatus 15 such as a ball mill. The powder 12 preferably has a size of less than 0.5 mm.

Examples of biomass are summarized in Table 11.4. During pyrolysis, the refinery feedstock may act as a hydrogen donor and in situ hydrogen deoxygenation of the biomass products may occur.

The powder 12 is then fed to an amount of refinery feedstock 16 contained within pyrolysis unit 18. A pyrolysis reactor has been described [37, 38].

The pyrolysis unit 18 includes a reactor 20, a heating means such as a combustion gas jacket 21, and a stirrer 22. The pyrolysis unit may further be provided with a condenser 29. Pyrolysis is generally conducted

Table 11.4 Examples of biomass

Biomass type		
Trees	Shrubs	Sawdust
Bark	Wood products	Municipal solid waste
Paper	Cardboard	Food waste
Leaves	Grass clippings	Plastics
Tires	Livestock waste	Organic process waste
Sewage	Starch crops	Sugar crops
Corn	Wheat	Barley
Cane	Beets	Grasses
Alfalfa	Clover	Switchgrass
Soybean crops	Sunflower crops	Safflower crops
Straw	Husks	Algae
Water weed	Water hyacinth	Reeds

Source: Jess A, Kaufmann D, Daugaard DE. Biomass pyrolysis in refinery feedstock. US Patent 8 287 723, assigned to Phillips 66 Company (Houston, TX); 2012. http://www.freepatentsonline.com/8287723.html.

in the absence of oxygen. The temperature of the pyrolysis unit is preferably from about 400 to 500°C. At lower temperatures, limited or no biomass conversion may occur, and at higher temperatures, there may be limited liquid production with an increase in noncombustible gas and coke production. The pressure of the pyrolysis unit 18 preferably ranges from atmospheric pressure to 20 MPa. At lower pressures, chemical interaction of hydrocarbon and biomass may be limited, and at higher pressures, chemical interaction of hydrocarbon and biomass may be promoted [36].

On initial contact between the powder 12 and the refinery feedstock 16, the biomass undergoes devolatilization, generating solids, such as char, condensable vapors, and noncondensable gases. The noncondensable gases may include CO, CO_2, CH_4, H_2, and O_2. The gases can be utilized to support the pyrolysis temperature by delivery to a combustion unit 28 through the line 30 or can be recovered for future use [36].

Other general aspects of the construction of pyrolysis units have been described in detail [39, 40]. A continuous flow-type pyrolysis reactor system has been described for the conversion of carbonizable input material. A diffusion material is used that is preheated and admixed with the input material to facilitate the pyrolysis and also to partially catalyze the pyrolysis reactions [40]. The system is substantially self-powered as a result of its using combustible hydrocarbon gases produced in the course of the pyrolysis reaction as fuel for turbulent burners. The input materials are evacuated to prevent the presence of oxygen. The pyrolysis system has been particularly adapted for use with tire chips, medical waste, and industrial plastic waste input materials. Carbon black and pelletized metallic and alloy materials are the preferred diffusion materials [40].

11.9 PRETREATMENT OF HEAVY OILS

The availability of light sweet crude oils is expected to diminish in the future as the production of oil becomes increasingly difficult and greater reliance is placed on tertiary and other enhanced recovery techniques [41].

Heavier crude oils and sour crude oils will become more important in the overall hydrocarbon production and the upgrading of such crude oils to fuels will become increasingly important.

Prerefining of crude oil could provide a significant advantage for downstream process units. The removal of metals as well as the reduction of aromatics and the removal of sulfur would substantially improve the performance of the subsequent coking units.

Heavy oils, including crude oils, vacuum residue, tar sands, bitumen, and vacuum gas oils, can be subjected to a pretreatment using a catalytic hydrotreating process [41]. Hydrodemetallization catalysts and hydrodesulfurization catalysts are used in series to improve the efficiency of a subsequent coker refinery.

One of the main limiting factors for hydrotreating units is the deactivation of the catalyst. Cobalt can be used to increase desulfurization of a hydrodesulfurization catalyst. The hydrodesulfurization catalyst may have a higher metal loading for the active phase in comparison with the hydrodemetallization catalyst. This increased metal loading helps to get an increased activity. Preferably, the hydrodesulfurization catalyst has a nickel-to-molybdenum molar ratio (Ni/(Ni+Mo)) of 0.1–0.3. When in addition cobalt is used, the (Co+Ni)/Mo molar ratio can be 0.25–0.85 [41].

The catalyst preferably has a mesoporous structure with pore sizes of 12–50 nm. This characteristic provides a balanced activity.

11.10 RECOVERY OF AROMATICS

The recovery of aromatics from gasoline fractions and refinery streams is an important step in refinery technology. Benzene and derivatives of benzene are important raw materials for the production of dyestuffs, plastics, solvents, and varnishes.

Reformate gasoline, pyrolysis gasoline, and distillation fractions from mineral oils or coke-oven light oil are examples of materials that contain aromatics.

It is not possible to separate aromatics from gasoline fractions containing aromatics by simple distillation, because the gasoline consists of a large variety of substances of very similar boiling ranges. Therefore the separation must be done by superior processes. These are [42]:

- azeotropic distillation,
- liquid-liquid extraction, and
- extractive distillation.

11.10.1 Azeotropic Distillation

In azeotropic distillation, a solvent is added to the mixture to be separated. The solvent forms a mixture of a constant boiling point with the aliphatic or the aromatic component. This azeotrope is separated from the starting mixture by distillation and after distillation is fractionated into the azeotrope and an aromatic fraction.

11.10.2 Liquid-Liquid Extraction

In liquid-liquid extraction, the mixture to be separated is provided with a solvent, generating a two-phase mixture and involving a higher solubility of one component and thus it is extracted selectively from the solvent mixture. The aromatic component may be separated from the solvent by distillation after the extraction [42].

11.10.3 Extractive Distillation

Extractive distillation has the advantage that there is a change in the fugacities in a mixture of appropriate components [42]. The fugacity is the corrected partial vapor pressure in the mixture. The change in the fugacity is caused there being different repellent interactions between the individual types of the molecules. A component in the mixture that has stronger repulsive forces than the other components will move into the vapor phase more easily than a component with lower repulsive forces.

In extractive distillation a solvent is added which is known to be capable of selectively increasing the fugacity of one or several components. In the case of hydrocarbon mixtures containing aromatics, the aliphatic components of the mixture frequently have stronger repulsive forces vis-à-vis the solvent, so their fugacity is considerably increased. In contrast, the fugacity of the aromatic component changes comparatively less. For this reason, a distillation with the solvent will result in the aliphatic components being preferably obtained in the raffinate, the low-boiling head product of a distillation, whereas the aromatic components are obtained in the extract, the higher-boiling bottom product of a distillation. This makes it necessary to use a solvent that achieves the desired effect by changing the fugacities of the individual components in the desired way [42].

Extractive distillation usually has advantages over azeotropic distillation or liquid-liquid extraction. The mass transfer in extractive distillation is frequently considerably higher than in azeotropic distillation. Extractive distillation requires considerably less equipment than liquid-liquid extraction, as only two distillation columns are required instead of one extraction column with a downstream distillation unit. Also, considerably less solvent is required for extractive distillation in comparison with liquid-liquid extraction.

The selectivity indicates the improvement of the extraction of the desired transition component in comparison with the other components contained in the raffinate. The higher the selectivity of an extracting agent, the stronger is the repulsion of the aliphatic component and thus the corresponding change in fugacity. The selectivity essentially governs the separating efficiency.

The actual equipment for an extractive distillation unit usually consists of two distillation columns. The first column serves to perform the actual extractive distillation. At the head of the column a raffinate stream is obtained which predominantly consists of nonaromatic hydrocarbons and, depending on the configuration, a certain amount of solvent. As the repulsive effect of the solvent is stronger for the nonaromatic hydrocarbons, these compounds change to the vapor phase more easily. At the lower part of the column a mixture is obtained which predominantly consists of aromatic compounds and the extracting solvent. This mixture is then passed into a stripping column where the aromatics-containing mixture is separated from the solvent by way of distillation. The solvent is recycled to the first column [42].

Various solvents suited for extractive distillation of aromatics are known and are summarized in Table 11.5. Recently, it was found that the two solvent additives and in combination with *N*-formylmorpholine are particularly well suited for the extractive distillation of aromatics [42]. These compounds are shown in Fig. 11.2.

Table 11.5 Solvents for extractive distillation of aromatics

Solvent or combination
Diethylene glycol
Dimethyl sulfoxide
Sulfolane
N-Methylpyrrolidone
Dimethylacetamide
N-Formylmorpholine
N-Formylmorpholine and *N*,*N*'-bis(formyl)piperazine
N-Formylmorpholine and 2,2'-bis(cyanoethyl)ether

Source: Noll O, Gehrke H, Luebbecke C, Kolbe B. Recovery of benzene and benzene derivatives from gasoline fraction and refinery streams. US Patent 8 536 397, assigned to UHDE GmbH (Dortmund, DE); 2013. http://www.freepatentsonline.com/8536397.html.

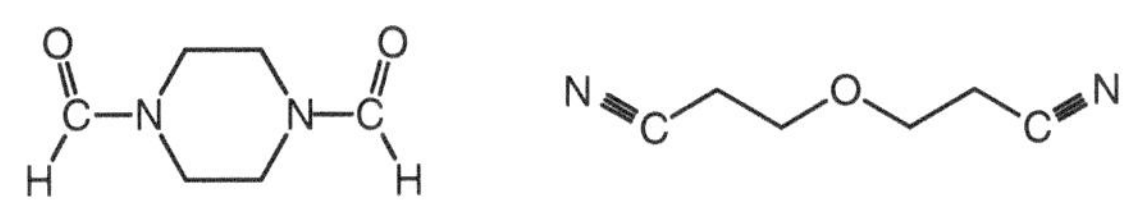

N.N´-Bis(formyl)piperazine 2.2´-Bis(cyanoethyl)ether

Figure 11.2 Solvent additives.

The capacity of these solvent combinations is so high that the aromatics can be separated from the gasoline fraction with only a small amount of extracting agent. The selectivity in relation to aromatics is so high that, after addition of the extracting solvent, the boiling points of the aromatic compounds differ sufficiently enough for a distillative separation. Thus the nonaromatic hydrocarbons are almost completely separated from the gasoline fraction, so after the removal of the extracting solvent it will be possible to perform a simple distillative separation of the aromatics benzene, toluene, and the xylene isomers [42].

The effect of the solvent combination is also supported by theoretical vapor liquid equilibrium data-based calculations. The process was simulated using the main plant parameters entered into the Aspen Plus computer program. Input data were temperature, pressure, boiling point, interaction parameters, and solubility of all specified compounds. Both solvent combinations were specified with a mass ratio of 1:1 in the simulation calculation.

11.11 RECOVERY OF HYDROGEN

Hydrogen can be used within hydrotreaters to remove sulfur, oxygen, and nitrogen and also within hydrocrackers to produce lighter and more paraffinic oils. Consequently, refineries look for low-cost sources of hydrogen [43].

While the refinery gases are a potential source of hydrogen, many refinery gas streams are not used either for their hydrogen content or to generate hydrogen through known reforming techniques. Often, the economics of separating hydrogen from refinery gases that contain less than about 30% hydrogen are unfavorable. Several processes are known for the recovery of hydrogen from hydrogen-hydrocarbon feed gas streams.

The problem of refinery hydrogen distribution has been elucidated [44]. A systematic method for the analysis of hydrogen distribution systems based on the concept of hydrogen surplus has been proposed. This method sets targets for the minimum flow rate of fresh hydrogen required by the refinery before any system design. The analysis method is also used to provide quantitative insights and to identify the existence of bottlenecks in a hydrogen distribution system.

11.11.1 Olefin Recovery and Hydrogen Production

An integrated method for olefin recovery and hydrogen production from a refinery off-gas has been described [45]. After a conventional pretreatment, the refinery gas is separated to obtain a light ends stream containing hydrogen, nitrogen, and methane and a heavy ends stream containing olefins. The light ends stream is mixed with natural gas and subjected to reforming and water gas shift reactions for hydrogen production. Further, the heavy ends stream can be processed for recovery of olefins, such as ethylene and propylene.

11.11.2 Pressure Swing System

A pressure swing cycle includes a feed step, a depressurization step, a purge step, and a repressurization step to prepare the adsorbent material for reintroduction of the feed [46]. The sorption of the contaminants usually occurs by physical adsorption. This consists typically of a porous solid such as alumina, silica, or silica-alumina. Zeolites can be used in many applications, because they may exhibit a significant degree of selectivity for certain contaminants.

Other types of cycling adsorption units can be used instead of a pressure swing adsorption (PSA) unit. For example, a temperature swing adsorption unit can operate in a manner similar to a PSA unit. Instead of using pressure as the primary driver for sorption and desorption of contaminants, a temperature swing adsorption process can use temperature as the primary variable for controlling contaminant sorption and desorption. Pressure changes can also be used in a temperature swing adsorption unit to further control contaminant removal, and similarly temperature changes can be used in a PSA unit [46].

A combination of a PSA process and a cryogenic process has been described for the recovery of light gases, such as hydrogen and helium [47, 48]. A method for separating hydrogen from impurities uses both a PSA unit and a membrane separator [49]. The initial hydrogen-containing feed is passed through the membrane to produce a hydrogen-enriched permeate and a retentate. The hydrogen-enriched permeate is used as the feed gas to the PSA for production of a hydrogen product stream. When the PSA unit has reached the end of an adsorption step, the retentate from the membrane is fed through the adsorption unit to help sweep out the hydrogen product that is still in the unit.

Another process and system for recovery of high-purity hydrogen and also hydrocarbon-rich products from a feed gas containing less than 50 mol% hydrogen via partial condensation in an autorefrigeration cold box and a PSA unit has been described [43]. This process can be used alone, only for hydrogen recovery, but can be also modified to allow the recovery of other valuable by-products such as methane-rich gas and raw methane-depleted gas containing heavier hydrocarbons. In the stand-alone process, the feed is compressed, treated for removal of water and impurities, cooled in a cold box, and then sent to a separator [43].

11.11.3 Membrane Separation Unit

A refinery network that includes both sources and consumers of hydrogen can generate at least two streams containing hydrogen whose recovery and reuse is desirable. Each of these streams can be available over a range of pressures.

The higher-pressure effluent can be passed to a membrane separation unit. The lower-pressure effluent can be sent to a cycling adsorption unit, such as a PSA unit, or a temperature swing adsorption unit. The membrane separation unit can be operated to produce a hydrogen-rich product of about 99.0% purity by volume. Subsequently, the hydrogen can be used as a low-pressure purge for a cycling adsorption unit [46]. This can reduce the portion of the hydrogen product stream from the cycling adsorption unit required for regeneration of the adsorbent [46].

11.12 CRACKING

Light olefins such as ethylene and propylene have traditionally been manufactured by cracking of hydrocarbon streams [50].

Gas oils typically have a final boiling point of up to 340°C and are typically derived from an atmospheric pipestill sidestream located just above the bottoms product. The atmospheric still bottoms product is commonly called *atmospheric resid* or *long resid*. Atmospheric resid can be provided to a vacuum pipestill operating at lower hydrocarbon partial pressures. The nonbottoms products of a vacuum pipestill are called *vacuum gas oils* and typically have a final boiling point of up to 650°C. The bottoms product of a vacuum still is known as vacuum resid, *short resid*, or *pitch*.

In general, cracking is the heating of hydrocarbon streams in the presence of steam up to 370°C. There, many of the hydrocarbons undergo cracking. As a result of the breaking of carbon-carbon bonds, hydrogen is released and also olefinic and aromatic products are formed [50].

11.12.1 High Boiling Point Hydrocarbons

An undesirable side reaction is the formation of coke deposits. The formation of coke causes an unacceptable increase in the pressure drop of the furnace and a loss of heat transfer. Methods to reduce coke formation have been developed [50]. One such method involves introduction of a flash operation within the convection sections of the pyrolysis furnace, where a heavy feedstock and steam mixture is preheated and separated. The overhead flash vapor of the heated mixture is then further heated to a higher temperature in the radiant section such that cracking occurs and olefins are produced. The flash operation produces a bottoms liquid product containing most of the problematic higher boiling point components that are not cracked.

A special method for cracking of heavy feedstocks has been developed. The methods consists in [50]:

1. feeding a hydrocarbon feedstock containing at least 1% of resid fractions having boiling points above 500°C to a furnace convection section to heat the feedstock,

2. flashing the heated feedstock in a first separation vessel to create a first overhead stream and a first bottoms liquid stream,
3. hydrogenating at least a portion of the first bottoms liquid stream to create a hydrogenated bottoms stream,
4. flashing the hydrogenated bottoms stream in a second separation vessel to create a second overhead stream and a second bottoms liquid stream, and
5. cracking the first overhead stream and the second overhead stream in a cracking furnace to produce a pyrolysis effluent stream.

11.12.2 Secondary Cracking

The vacuum residuum was thermally treated in a microbatch reactor at 410–480°C [51]. The variation of the products with time indicated that a secondary cracking mainly occurs in the heavy fraction of the liquid product.

A narrow fraction model was developed for low-severity thermal cracking of heavy oil that takes into account also the secondary cracking reactions. A sequential method to estimate the various rate constants was used. The kinetic parameters were estimated by the Arrhenius law. The yields predicted by this model agreed well with the experimental results. Also, the results indicated that the secondary cracking reaction cannot be ignored even in low-severity thermal cracking [51].

11.13 COKING

There are three main variants of the coking process: delayed coking, fluid coking, and its variant, the Flexicoking™ process [52]. Delayed coking is the variant with the greatest global installed capacity, while the fluid coking processes (coking and Flexicoking) are becoming of greater interest with the need to maximize efficiency and utilize resources as efficiently as possible. The moving bed continuous contact coking process now appears to be obsolete.

Fluidized bed coking is a petroleum refining process in which heavy petroleum feeds, typically the nondistillable residue (resid) from fractionation, are converted to lighter, more useful products by thermal decomposition (coking) at elevated reaction temperatures, typically about 480–590°C. The process is performed in a unit with a large reactor vessel containing hot coke particles which are maintained in the fluidized condition at the required reaction temperature with steam injected at the bottom of the vessel with the average direction of movement of the coke particles being downward through the bed. The heavy oil feed is heated to a pumpable temperature, mixed with atomizing steam, and fed through multiple feed nozzles arranged at several successive levels in the reactor. The steam is injected into a stripper section at the bottom of the reactor and passes upward through the coke particles in the stripper as they descend from the main part of the reactor above. Part of the feed liquid coats the coke particles in the fluidized bed and subsequently decomposes into layers of solid coke and lighter products, which evolve as gas or vaporized liquid. The light hydrocarbon products of the coking (thermal cracking) reactions vaporize, mix with the fluidizing steam, and pass upward through the fluidized bed into a dilute phase zone above the dense fluidized bed of coke particles. This mixture of vaporized hydrocarbon products formed in the coking reactions continues to flow upward through the dilute phase with the steam at superficial velocities of about 1–2 ms^{-1}, entraining some fine solid particles of coke. Most of the entrained solids are separated from the gas phase by centrifugal force in one or more cyclone separators, and are returned to the dense fluidized bed by gravity through the cyclone diplegs. The mixture of steam and hydrocarbon vapor from the reactor is subsequently discharged from the cyclone gas outlets into a scrubber section in a plenum located above the reaction section. The heavy fraction separated in the scrubber is typically recycled to extinction by its being fed back to the fluidized bed reaction zone while the lighter cracking products pass on to the product recovery section.

The Flexicoking process, developed by Exxon Research and Engineering Company, is, in fact, a fluidized bed coking process that is operated in a unit including a reactor and burner, often referred to as a heater, also including a gasifier for gasifying the coke product by reaction with an air-steam mixture to form a low heating value fuel gas. The heater, in this case, is operated with an oxygen-depleted environment. The gasifier product gas, containing entrained coke particles, is returned to the heater to provide a portion of the reactor heat requirement. A return stream of coke sent from the gasifier to the heater provides the remainder of the heat requirement. Hot coke gas leaving the heater is used to generate high-pressure steam before being processed for cleanup. The coke product is continuously removed from the reactor. In view of the similarity between the Flexicoking process and the fluid coking process, the term *fluid coking* may be used for both processes to refer to and understand both fluid coking and Flexicoking except when a differentiation is required [52].

Fluid coking and Flexicoking technologies have been reviewed [53]. Flexicoking is a process that allows the conversion of the bottom separated material of the crude barrel to clean products [54]. Thus Flexicoking is an integrated coking and gasification process for upgrading heavy feedstocks. This process converts these feeds to a 99% yield of fuel gas, naphtha, middle distillates, heavy gas oil, and a low-sulfur coke gas. The remaining 1% is petroleum coke containing metals and other ash components present in the feed.

The enhancement of liquid yields from both processes and the reduction of low-Btu gas from Flexicoking have been discussed, as have the recent advances in coking technology [54].

11.14 OLEFIN AND REFORMATE PRODUCTION

Steam cracking is the thermal cracking of hydrocarbons in the presence of steam [55]. This method is used commercially in large-scale industrial units to produce ethylene and propylene. These pyrolysis units are often charged with a naphtha boiling range feed stream. The typical petroleum-derived naphtha contains a wide variety of different hydrocarbon types, including normal paraffins, branched paraffins, olefins, naphthenes, benzene, and alkyl aromatics. Paraffins are most easily cracked and provide the highest yield of ethylene. Some compounds, such as benzene, are relatively refractory to the typical cracking conditions.

Cracking normal paraffins results in a higher product yield than cracking isoparaffins [55]. The ethylene yield of a cracking unit can be increased if it is rather charged with a C_5-C_9 stream of normal paraffins instead of a typical natural gasoline with the same molecular distribution [56].

11.14.1 Paraffin Separation

To separate normal from nonnormal paraffins in naphtha feedstocks, adsorptive processes have been proposed. These processes usually use molecular sieves [57]. Unfortunately, many of the sorbents used exhibit a low tolerance for sulfur-, nitrogen-, and oxygen-containing components.

Thin zeolite membranes with an *n*-butane permeability of 6×10^{-7} $\mathrm{mol\,m^{-2}\,s\,Pa}$ and an *n*-butane selectivity of at least 250 have been described [58].

Sieving membranes can be used to separate a naphtha feedstock into a retentate fraction having a reduced concentration of normal paraffins [55]. A sieving membrane is a composite membrane containing a continuous or discontinuous selective separation medium of a molecular sieve barrier. A barrier is the structure that exists to selectively block fluid flow in the membrane.

In a continuous sieving membrane, the molecular sieve itself forms a continuous layer that is sought to be defect-free. The continuous barrier may contain other materials such as would be the case with mixed matrix membranes.

A discontinuous sieving membrane is a discontinuous assembly of a molecular sieve barrier in which spaces, or voids, exist between particles or regions of the molecular sieve. These spaces or voids may contain or be filled with other solid material. The particles or regions of the molecular sieve are the barrier.

The separation effected by sieving membranes may be on the basis of the steric properties of the components to be separated. Other factors may also affect permeation. One is the sorptivity or lack thereof of a component and the material of the molecular sieve. Another is the interaction of components to be separated in the microporous structure of the molecular sieve.

For some zeolitic molecular sieves, the presence of a molecule (eg, *n*-hexane) in a pore may hinder 2-methylpentane from entering that pore more than another *n*-hexane molecule [55].

Polymeric molecular sieves can be synthesized with both a mesoporous structure and a macroporous structure. One method for synthesizing a small polymeric molecular sieve is to functionalize nanoparticles or the mesoporous/macroporous structure with a group that can react with an oligomer such as through a condensation or addition reaction. The functional groups may contain a hydroxyl, amino, anhydride, dianhydride, aldehyde, carboxyl, amide, nitrile, or olefinic moiety suitable for an addition or condensation reaction with a reactive moiety of an oligomer [55].

11.14.2 Methyl *tert*-Butyl Ether Bypass

Methyl *tert*-butyl ether (MTBE) is manufactured by the chemical reaction of methanol and isobutene for primary use in gasoline [59]. MTBE is a common component in reformulated fuels developed to reduce smog and meet the goals of the Clean Air Act. MTBE has been produced in very large quantities for use as a gasoline additive since about 1979.

However, MTBE production has decreased as various jurisdictions restricted or banned its use. By late 2006 most American gasoline retailers stopped using MTBE as an oxygenate. Accordingly, the domestic production has continued to decline. As a result, MTBE manufacturers are left holding useless feedstocks and manufacturing assets.

Therefore it is desirable to reallocate the feedstocks and manufacturing assets previously allocated to the manufacture of MTBE, thereby providing an economic benefit to MTBE manufacturers [59].

A process for producing olefins via the MTBE process has been described. A feed containing C_4 compounds can be dehydrogenated to yield butene. At least a portion of this product can be bypassed around an MTBE unit and cracked, resulting in propylene, ethylene, and butane. Light hydrocarbons can be cracked in another cracker to give again propylene, ethylene, and butane [59].

REFERENCES

[1] Gary J, Handwerk GE, Petroleum refining: technology and economics. New York: Marcel Dekker; 1984. ISBN 0824771508.

[2] Speight J, Petroleum chemistry and refining. Washington, DC: Taylor & Francis; 1998. ISBN 1560325879.

[3] Predel H, Petroleum coke. In: Ullmann's encyclopedia of industrial chemistry. Weinheim: Wiley-VCH Verlag GmbH & Co. KGaA; 2000. http://id.crossref.org/isbn/9783527306732, doi:10.1002/14356007.a19_235.pub2.

[4] Parkash S, Refining processes handbook. Amsterdam/Boston: Gulf Professional Pub; 2003. ISBN 075067721X.

[5] Fahim MA, Al-Sahaf TA, Elkilani AS, Fundamentals of petroleum refining. Amsterdam/London: Elsevier Science; 2010. ISBN 0080931561.

[6] Butler G, Couves JW, Greenough P, Gudde NJ, Hodges MG. Process for evaluating a refinery feedstock. US Patent 8 546 146, assigned to BP Oil International Limited (Middlesex, GB); 2013. http://www.freepatentsonline.com/8546146.html.

[7] Kremer LN, Nguyen TM, Weers JJ. Additives to enhance phosphorus compound removal in refinery desalting processes. US Patent 7 799 213, assigned to Baker Hughes Incorporated (Houston, TX); 2010. http://www.freepatentsonline.com/7799213.html.

[8] Lindemuth PM. Abatement of hydrolyzable cations in crude oil. US Patent 5 660 717, assigned to Nalco/Exxon Energy Chemicals, L. P. (Sugarland, TX); 1997. http://www.freepatentsonline.com/5660717.html.

[9] noz MJGM, Rodriguez MA, Luque S, Álvarez JR, Recovery of heavy metals from metal industry waste waters by chemical precipitation and nanofiltration. Desalination 2006;200(1–3):742–4. http://www.sciencedirect.com/science/article/pii/S0011916406006230, http://dx.doi.org/10.1016/j.desal.2006.03.498.

[10] Tomita T, Nakayama K, Sakai H, Gas separation characteristics of {DDR} type zeolite membrane. Microporous Mesoporous Mater 2004;68(1–3):71–5. http://www.sciencedirect.com/science/article/pii/S1387181103006772, http://dx.doi.org/10.1016/j.micromeso.2003.11.016.

[11] Nguyen TM, Kremer LN, Weers JJ. Additives to enhance metal and amine removal in refinery desalting processes. US Patent 8 372 271, assigned to Baker Hughes Incorporated (Houston, TX); 2013. http://www.freepatentsonline.com/8372271.html.

[12] Nguyen TM, Weers JJ, Yang J, Babic K. Additives to enhance metal removal in refinery desalting processes. US Patent 8 372 270, assigned to Baker Hughes Incorporated (Houston, TX); 2013. http://www.freepatentsonline.com/8372270.html.

[13] Wang FC, Wright CA, Brons GB, Levine SW, Farng LO. Metal sulphonate additives for fouling mitigation in petroleum refinery processes. WO Patent 2010 019 543 assigned to Exxonmobil Research And Engineering Company; 2010. https://www.google.at/patents/WO2010019543A2?cl=en.

[14] Greaney MA, Bruno T, Cooper AE, Cody IA, Chun C. Corrosion resistant material for reduced fouling, a heat transfer component having reduced fouling and a method for reducing fouling in a refinery. US Patent 8 469 081, assigned to ExxonMobil Research and Engineering Company (Annandale, NJ); 2013. http://www.freepatentsonline.com/8469081.html.

[15] Park CM, Richardson EE. Preparation of highly based magnesium sulfonate. US Patent 4 474 710, assigned to Standard Oil Company (Indiana) (Chicago, IL); 1984. http://www.freepatentsonline.com/4474710.html.

[16] Budhram RS, Eisenbraun EJ, Synthesis and rigorous purification of sodium alkylbenzene sulfonates. J Am Oil Chem Soc 1986;63(10):1392–4. http://dx.doi.org/10.1007/BF02679609.

[17] Yang J, Qiao W, Li Z, Cheng L, Synthesis and characterization of mono-isomeric alkylbenzene sulfonates. Pet Sci Technol 2006;24(8):973–84. http://dx.doi.org/10.1081/LFT-200048161.

[18] Totten GE, Handbook of hydraulic fluid technology. Boca Raton, FL: CRC Press; 1999.

[19] Forsberg JW. Aqueous systems containing nitrogen-containing, phosphorous-free carboxylic solubilizer/surfactant additives. US Patent 4 368 133, assigned to The Lubrizol Corporation (Cleveland, OH); 1983. http://www.freepatentsonline.com/4368133.html.

[20] Brons G, Brown LD, Joshi H, Kennedy RJ, Bruno T, Rudy TM. Method for refinery foulant deposit characterization. US Patent 7 708 864, assigned to ExxonMobil Research & Engineering Company (Annandale, NJ); 2010. http://www.freepatentsonline.com/7708864.html.

[21] ASTM D5373. Standard test methods for determination of carbon, hydrogen and nitrogen in analysis samples of coal and carbon in analysis samples of coal and coke. ASTM Standard, Book of Standards, Vol. 05.06 ASTM D5373, ASTM International, West Conshohocken, PA; 2014.

[22] Yeganeh MS, Jacobs PW, Cody IA, Levine SW, Sirota EB, Gupta R, et al. Method and systems to remove polar molecules from refinery streams. US Patent 8 852 427, assigned to ExxonMobil Research and Engineering Company (Annandale, NJ); 2014. http://www.freepatentsonline.com/8852427.html.

[23] Logan MJ, Pearce PR, Dick DG. Production of a refinery feedstock from soaps produced during a chemical pulping process. US Patent 7 540 889, assigned to Bluekey Energy Inc. (Prince George, BC, CA); 2009. http://www.freepatentsonline.com/7540889.html.

[24] Chen CY, Miller SJ, Ziemer JN. Multi-stage reforming process to produce high octane gasoline. US Patent 8 882 992, assigned to Chevron U.S.A. Inc. (San Ramon, CA); 2014. http://www.freepatentsonline.com/8882992.html.

[25] Bourane A, Koseoglu O. Catalytic reforming process and system for producing reduced benzene gasoline. US Patent 8 801 920, assigned to Saudi Arabian Oil Company (Dhahran, SA); 2014. http://www.freepatentsonline.com/8801920.html.

[26] Oyekan SO, Rhodes KD, Newlon NK. Feed additives for CCR reforming. US Patent 8 778 823, assigned to Marathon Petroleum Company LP (Findlay, OH); 2014. http://www.freepatentsonline.com/8778823.html.

[27] Peng XD, Achilles GC, Herb BE, Klingenberg EH, Miller D. Water purification using energy from a steam-hydrocarbon reforming process. US Patent 8 709 287, assigned to Air Products and Chemicals, Inc. (Allentown, PA); 2014. http://www.freepatentsonline.com/8709287.html.
[28] Brown SH, Weber WA, Shinnar R. Process for converting natural gas to higher value products using a methanol refinery remote from the natural gas source. US Patent 6 632 971, assigned to ExxonMobil Chemical Patents Inc. (Houston, TX); 2003. http://www.freepatentsonline.com/6632971.html.
[29] Chang CD, Lang WH, Silvestri AJ. Conversion of methanol to olefinic components. US Patent 4 052 479, assigned to Mobil Oil Corporation (New York, NY); 1977. http://www.freepatentsonline.com/4052479.html.
[30] Brown SH, Green LA, Mathias MF, Olson DH, Ware RA, Weber WA. Process for producing light olefins. US Patent 6 046 372, assigned to Mobil Oil Corporation (Fairfax, VA); 2000. http://www.freepatentsonline.com/6046372.html.
[31] Seddon D, Mole T, Whiteside JA. Methanol conversion to hydrocarbons with zeolites and cocatalysts. US Patent 4 499 314, assigned to Imperial Chemical Industries PLC (London, GB2); 1985. http://www.freepatentsonline.com/4499314.html.
[32] Barger PT, Wilson ST, Holmgren JS. Metal aluminophosphate catalyst for converting methanol to light olefins. US Patent 5 126 308, assigned to UOP (Des Plaines, IL); 1992. http://www.freepatentsonline.com/5126308.html.
[33] Barger PT, Wilson ST, Holmgren JS. Process for converting methanol to olefins using an improved metal aluminophosphate catalyst. US Patent 5 191 141, assigned to UOP (Des Plaines, IL); 1993. http://www.freepatentsonline.com/5191141.html.
[34] Schinski WL. Hybrid refinery for co-processing biomass with conventional refinery streams. US Patent 8 641 991, assigned to Chevron U.S.A. Inc. (San Ramon, CA); 2014. http://www.freepatentsonline.com/8641991.html.
[35] Gomes JR, Zotin JL, Pacheco ME, Cotta Ferreira Da Silva RM. Process for hydrotreating biomass oil diluted in a refinery stream of petroleum hydrocarbons. US Patent 8 507 738, assigned to Petroleo Brasileiro S.A.—Petrobras (Rio de Janeiro, BR); 2013. http://www.freepatentsonline.com/8507738.html.
[36] Jess A, Kaufmann D, Daugaard DE. Biomass pyrolysis in refinery feedstock. US Patent 8 287 723, assigned to Phillips 66 Company (Houston, TX); 2012. http://www.freepatentsonline.com/8287723.html.
[37] Keusenkothen PF, McCoy JN, Hershkowitz F. Pyrolysis reactor conversion of hydrocarbon feedstocks into higher value hydrocarbons. US Patent 8 303 803, assigned to ExxonMobil Chemical Patents Inc. (Houston, TX); 2012. http://www.freepatentsonline.com/8303803.html.
[38] Venderbosch RH, Van De Beld L, Assink D, Gansekoele E. Pyrolysis reactor. US Patent 8 585 868, assigned to BTG BioLiquids B.V. (Enschede, NL); 2013. http://www.freepatentsonline.com/8585868.html.
[39] Chittick DE. Fuel gas-producing pyrolysis reactors. US Patent 4 584 947; 1986. http://www.freepatentsonline.com/4584947.html.
[40] Wu AC, Chen SC. Thermal conversion pyrolysis reactor system. US Patent 5 411 714; 1995. http://www.freepatentsonline.com/5411714.html.
[41] Shafi R, Hamad EZ, Kressmann SC, Moses JH. Alternative process for treatment of heavy crudes in a coking refinery. US Patent 8 491 779, assigned to Saudi Arabian Oil Company (SA); 2013. http://www.freepatentsonline.com/8491779.html.
[42] Noll O, Gehrke H, Luebbecke C, Kolbe B. Recovery of benzene and benzene derivatives from gasoline fraction and refinery streams. US Patent 8 536 397, assigned to UHDE GmbH (Dortmund, DE); 2013. http://www.freepatentsonline.com/8536397.html.

[43] Dragomir RM, Drnevich RF, Bonaquist DP. Refinery gas upgrading via partial condensation and PSA. US Patent 8 535 415, assigned to Praxair Technology, Inc. (Danbury, CT); 2013. http://www.freepatentsonline.com/8535415.html.
[44] Alves JJ, Towler GP, Analysis of refinery hydrogen distribution systems. Ind Eng Chem Res 2002;41(23):5759–69. http://dx.doi.org/10.1021/ie010558v.
[45] Drnevich RF, Herzog JO. Integrated olefin recovery and hydrogen production from refinery off-gases. US Patent 7 041 271, assigned to Praxair Technology, Inc. (Danbury, CT); 2006. http://www.freepatentsonline.com/7041271.html.
[46] Sundaram N, Viets JW. Hydrogen utilization within a refinery network. US Patent 8 512 443, assigned to ExxonMobil Research and Engineering Company (Annandale, NJ); 2013. http://www.freepatentsonline.com/8512443.html.
[47] Meisler J, Banikiotes GC, Van Baush EH. Hydrogen purification. US Patent 3 691 779, assigned to Hydrocarbon Research, Inc. (New York, NY); 1972. http://www.freepatentsonline.com/3691779.html.
[48] Doherty KS. Separation of mixtures especially gas mixtures. US Patent 3 838 553, assigned to Petrocarbon Developments Limited (Manchester, EN); 1974. http://www.freepatentsonline.com/3838553.html.
[49] Doshi KJ. Enhanced hydrogen recovery from low purity gas streams. US Patent 4 398 926, assigned to Union Carbide Corporation (Danbury, CT); 1983. http://www.freepatentsonline.com/4398926.html.
[50] Kuechler KH, Bancroft JL, Keusenkothen PF, Strack RD. Process and apparatus for cracking high boiling point hydrocarbon feedstock. US Patent 8 882 991, assigned to ExxonMobil Chemical Patents Inc. (Houston, TX); 2014. http://www.freepatentsonline.com/8882991.html.
[51] Yang J, Chen D, Shen G, Xiao J, Yang C, Narrow fraction model with secondary cracking for low-severity thermal cracking of heavy oil. Energy Fuels 2012;26(6):3628–33. http://dx.doi.org/10.1021/ef3004625.
[52] Siskin M, Phillips GE, Kelemen SR, Weissman W. Biomass oil conversion process. US Patent 8 603 325, assigned to ExxonMobil Research and Engineering Company (Annandale, NJ); 2013. http://www.freepatentsonline.com/8603325.html.
[53] Hammond DG, Lampert LF, Mart CJ, Massenzio SF, Phillips GE, Sellards DL, et al. Review of fluid coking and flexicoking technologies. In: AIChE 6th Topical Conf. on Refining Processing, USA, Spring; 2003.
[54] Allan DE, Metrailer WJ, King RC, Wiechert S. Processing heavy crudes: advances in fluid and flexicoking technology. Chem Eng Prog 1981;77.
[55] Rice LH. Integrated refinery with enhanced olefin and reformate production. US Patent 7 846 322, assigned to UOP LLC (Des Plaines, IL); 2010. http://www.freepatentsonline.com/7846322.html.
[56] Reddock IA, Separation of normal paraffins from isoparaffins. Brisbane: Eleventh Australian Conference on Chemical Engineering; 1983.
[57] Foley TD, Sohn SW. Ethylene production by steam cracking of normal paraffins. US Patent 6 407 301, assigned to UOP LLC (Des Plaines, IL); 2002. http://www.freepatentsonline.com/6407301.html.
[58] Chau C, Sicard M, Le Dred R. Thin zeolite membrane, its preparation and its use in separation. US Patent 6 818 333, assigned to Institut Francais du Petrole (Rueil Malmaison Cedex, FR); 2004. http://www.freepatentsonline.com/6818333.html.
[59] Eng CN. Method for olefin production from butanes and cracking refinery hydrocarbons and alkanes. US Patent 8 080 698, assigned to Kellogg Brown & Root LLC (Houston, TX); 2011. http://www.freepatentsonline.com/8080698.html.

CHAPTER 12

Safety Aspects for Refineries

12.1 ESTIMATES OF EMISSIONS

Estimates of emissions for processes and point sources at petroleum refineries and chemical plants are necessary for other environmental evaluations and policy decisions. However, the most commonly used method is based on emission factors and results in unreliable estimates [1]. Therefore more information regarding the actual emissions is necessary to provide a foundation for improving the emission factors.

Benzene and volatile organic compound emissions within a major chemical refinery have been measured and compared with emission factor estimates. The results indicate that the estimated emissions are commonly lower than the measured emissions. The sources with the largest relative error were found to be a wastewater treatment process, an aromatics concentration unit, and a benzene extraction unit process area.

12.2 WASTE STREAM TREATMENT

12.2.1 Nitrogen Oxides

A fluid catalytic cracking unit has been described that contains an assembly that can convert nitrogen oxides into nitrogen [2]. The system contains a reactor unit for reacting a process stream with a catalyst, a regenerator unit for regenerating the catalyst, and a CO boiler unit for processing regenerator flue gas. The method of processing the waste stream includes the introduction of the waste stream into the CO boiler unit such that the waste stream and the regenerator flue gas are burned within the CO boiler unit. The waste stream contains NH_3, and the NH_3 is converted to NO_x and N_2 in the CO boiler unit.

Flue gas from the CO boiler unit containing the NO_x is fed to the assembly to convert NO_x to N_2 within the assembly. The assembly

Guide to the Practical Use of Chemicals in Refineries and Pipelines
http://dx.doi.org/10.1016/B978-0-12-805412-3.00012-X

preferably contains a selective catalytic reduction unit containing the catalyst [2]. As a catalyst, zeolites can be used.

12.2.2 Flocculation and Ceramic Membrane Filtration

Oily wastewater produced from refinery processes can be treated by flocculation and microfiltration with a zirconia membrane [3]. Studies revealed that the oil content and the chemical oxygen demand (COD) can be decreased dramatically by flocculation.

Also, the influence of the flocculation conditions on the flocculation results was investigated by orthogonal experiments. The optimum conditions are a dose of 70 $mg L^{-1}$, a temperature of 40°C, a stirring time of 90 min, and a holding time of 90 min.

After flocculation, the effluents were treated with a microfiltration unit using a zirconia membrane. The results of the filtration tests showed that the membrane fouling decreases and the permeate flux and permeate quality increase with flocculation as pretreatment [3].

12.2.3 Activated Carbon

The effects of powdered activated carbon on the performance of a dissolved air flotation unit were investigated [4]. Refinery wastewater of different pollutant concentrations was treated and the effects of different operating parameters on the removal efficiency of pollutants in terms of biological oxygen demand (BOD) and COD were studied.

It was found that for doses of activated carbon in the range of 50–150 $mg L^{-1}$, the removal efficiencies for BOD increased from 27–70% to 76–94%, while those for COD increased from 16–64% to 72–92.5% for inlet values of 45–95 $mg L^{-1}$ and 110–200 $mg L^{-1}$ for BOD and COD, respectively [4].

12.2.4 Demulsifiers in Oily Wastewaters

In industry, oily wastewaters are produced in various processes found in the steel and aluminum industries, the chemical processing industry, the automotive industry, the laundry industry, and the refinery industry. In these industries, highly refined oils, lubricants, and greases contact water. This may result in a highly dispersed oil-in-water emulsion in the wastewater streams.

Wastewater from steel and aluminum mills using hot rolling mills contain lubricating and hydraulic-pressure hydrocarbons. Wastewater from cold rolling mills contain oil that lubricates the sheets and reduces rust. For example, in cold rolling mills, oil-in-water emulsions are sprayed on the metal during rolling to act as coolants.

Metalworking plants generate wastewater streams containing lubricating and cutting oils, lapping and deburring compounds, and grinding and other specialty fluids. These oils are generally highly refined hydrocarbons. The wastewater streams in animal rendering plants, creameries, bakeries, breweries, and canneries contain natural fats and oils from animal processing and oils from packing and container manufacture. Wastewaters from cotton and wool manufacturing plants contain oils and greases from the scouring, desizing, and finishing operations. Finishing oils are used in cotton and wool manufacturing to reduce friction and snagging of fibers on spinning machines which use machine lubricating oils; these oils end up in the wastewater. Processes in other industries also generate oily wastewater, such as from paints, surface coatings, and adhesives, oils, fats, and waxes, soaps and detergents, dyes and inks, and the leather industry.

The emulsified oil in the wastewater is typically present in the range of several hundred to tens of thousands of parts per million. It is critical to remove this oil from an environmental standpoint. The United States Environmental Protection Agency has placed tight restrictions on total oil and grease limits for water that is to be discharged into public drinking water supplies or into open bodies of water. The removal of this oil is very critical to the established limits for discharge of total dissolved solids, COD, and total organic carbon into local sewers and rivers. Not only has the Environmental Protection Agency established severe limits on oil and grease discharge, the associated industries are also affected by local city ordinances as well.

One of the most effective methods of removing the emulsified oil in wastewater is through the use of the chemical aids. Typically, these dispersions or emulsions require 10–100 times the dose needed for other emulsion-breaking applications, such as removal of crude oil from water.

Historically, dry polymers, solution polymers, inverse emulsion latexes, and metal ions have been used to treat the water produced. Each material has its advantages and disadvantages. While dry polymers have the benefit of being extremely concentrated, thereby reducing shipping costs, the equipment to dissolve the polymers is expensive and is not available to all end users on-site.

Latex polymers also suffer from problems but are used quite frequently. Latex polymer preparations include 30–35% solids dispersed in oil. The latex polymer must be inverted before use. Numerous problems associated with this feeding method have caused many customers to avoid latex polymers. In addition, the latexes generally have a very narrow treatment range, often resulting in overtreatment at higher doses. Furthermore, latex polymers add even more oil to the stream to be treated. Of course, adding more oil is something most customers would not want to do when treating their wastewater streams.

Although solution polymers require no prior makeup, the percent solids and molecular weight are severely limited owing to the nature of the material. Such materials are often used to break oil-in-water emulsions, but they are unable to flocculate the dispersed oil, thus requiring another chemical to accomplish this.

Metal ions, such as Fe^{3+}, Zn^{2+}, and Al^{3+}, have long been used to break oil-in-water emulsions, but government regulations have restricted their levels in discharged streams. Although effective at breaking oil-in-water emulsions, they, too, require another chemical to flocculate the oil.

Vinylamine polymers offer many solutions to these problems and represent a new method for treatment of industrial wastewaters. Vinylamine polymers have been shown to be more effective in removing emulsified oil from industrial wastewaters than conventional chemical treatments. These polymers offer numerous advantages which have previously been unavailable to operators of industrial processes. The polymers are water soluble, and unlike with latex polymers, there is no oil solvent.

This is important because [5]:

- the polymers do not present a fire hazard,
- oil is not added to the water which is to be treated (ie, more environmentally friendly),
- the dissolution of the polymer requires only the addition of water—no special activators are needed,
- the ability of these materials to dissolve or invert is superior to that of oil dispersion latexes, and
- the polymers may be diluted with water to virtually any concentration.

These physical properties enable any standard chemical pump to deliver the material at the injection site. Processes for synthesizing such polymers have been described.

Solution polymerization produces the desired molecular weight dose. The resulting vinylformamide homopolymers are susceptible to alkaline or acid hydrolysis, which converts some or all of the amide groups to amine groups.

Table 12.1 Efficiency of NALCO® 8105 in emulsion breaking

Dose (ppm)	Turbidity (NTU)
0	293
1.9	120
3.7	713
5.6	55.4
7.4	43.8
9.3	41.1
11.1	40.6
13	41.7
14.8	39.8
16.7	83.2
18.5	110
20.4	139
22.2	156

NTU, nephelometric turbidity units.

Hydrolysis of the amide groups to yield the amine can be achieved by use of acid or base. By controlling the stoichiometry of the hydrolyzing agent one can produce vinylamine-vinylformamide copolymers with varying composition. If vinylformamide is polymerized with other ethylenically unsaturated monomers, then hydrolyzed, it is conceivable that the polymer produced will contain at least three and possibly more functional groups. Cyclic amidine is produced by the amine hydrolysis of polyvinylformamide.

NALCO® 7134 and NALCO® 8105 are solution polymers which are commonly used in refinery wastewater applications. The efficiency of NALCO® and other polymer compositions in emulsion breaking of a refinery wastewater is shown in Tables 12.1–12.3.

12.2.5 Wastewater Treatment

Membrane Technology

The development of membrane technology has attracted great attention in the fields of industrial water and wastewater treatment. It has been successfully implemented in many sectors, such as water desalination, ultra pure water production, product recycling, and wastewater reclamations. For refinery wastewater systems, the separation performance of a wettability microporous ultra filtration membrane depends mainly on the sieving effects of the pore structure [6] and the properties of the top surface [7–10].

Table 12.2 Efficiency of NALCO® 7134 in emulsion breaking

Dose (ppm)	Turbidity (NTU)
0	293
5	99.3
10	77.3
15	97.3
20	125
25	89.2
30	117
35	193
40	210
45	194

NTU, nephelometric turbidity units.
Source: Sommese AG, Sivakumar A. Polymers containing vinylamine/vinylformamide as demulsifiers in oily wastewaters. US Patent 5 702 613, assigned to Nalco Chemical Company (Naperville, IL); 1997. http://www.freepatentsonline.com/5702613.html.

Table 12.3 Efficiency of polyvinylamine plus 10% octyl chloride (30–50 kDa) in emulsion breaking

Dose (ppm)	Turbidity (NTU)
0	272
6	127
12	41
24	15
36	16
48	14
60	16
72	23
84	34
96	13

NTU, nephelometric turbidity units.
Source: Sommese AG, Sivakumar A. Polymers containing vinylamine/vinylformamide as demulsifiers in oily wastewaters. US Patent 5 702 613, assigned to Nalco Chemical Company (Naperville, IL); 1997. http://www.freepatentsonline.com/5702613.html.

The effect of additives on the surface properties and the performance of poly(vinylidene fluoride) (PVDF) ultrafiltration membranes for refinery produced wastewater treatment has been evaluated [11].

PVDF ultrafiltration membranes were prepared via a phase inversion method by dispersal of lithium chloride monohydrate and titanium dioxide nanoparticles in the spinning dope. The morphological and performance tests were conducted on PVDF ultrafiltration membranes prepared from a different additive content. The top surface and cross-sectional area of the membranes were observed by field emission scanning electron microscopy and energy-dispersive X-ray analysis.

The surface wettability of porous membranes was determined by the measurement of the contact angle. The mean pore size and surface porosity were calculated on the basis of the permeate flux. The results indicated that the membranes with a lower TiO_2 nanoparticle loading possessed a smaller mean pore size, more apertures inside the membrane, and enhanced membrane hydrophilicity.

$LiCl \cdot H_2O$ has been used particularly to reduce the thermodynamic miscibility of dope, which resulted in an increase in the rate of the liquid-liquid demixing process. The maximum flux and rejection of refinery wastewater when a PVDF ultrafiltration membrane was used achieved were 82.50 $L\,m^{-2}\,h^{-1}$ and 98.83%, respectively, at 1.95% TiO_2 concentration [11]. The process is shown in Fig. 12.1.

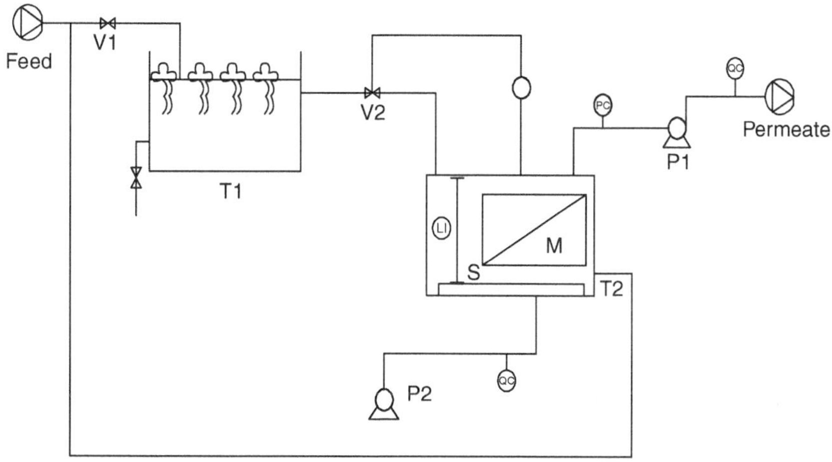

Figure 12.1 Submerged ultrafiltration process: *V*1 and *V*2, Feed flow; *P*1, peristaltic pump; *P*2, compressor; *S*, sparger; *M*, submerged membrane module; *T*1, pretreatment tank; *T*2, membrane reservoir; *QC*, flow control; *PC*, pressure control; *LC*, level control; *LI*, level indicator. *(Source: Yuliwati E, Ismail AF, Effect of additives concentration on the surface properties and performance of PVDF ultrafiltration membranes for refinery produced wastewater treatment. Desalination 2011;273(1):226–34. http://dx.doi.org/10.1016/j.desal.2010.11.023).*

Phenol

Phenol is classified as a priority pollutant by the US Environmental Protection Agency. The removal of phenol from a synthetic wastewater was experimentally investigated by use of different removal techniques, including [12]:

1. adsorption using granular activated carbon,
2. biodegradation using six different active bacteria that take up phenol,
3. combined granular activated carbon sorption and biodegradation,
4. granular activated carbon coupled with inactive bacteria (biosorption).

A solution of phenol at 100 $mg\,L^{-1}$ at room temperature was used to simulate the wastewater. The study was performed at 25–30°C. The results obtained with the different techniques revealed that activated carbon sorption is better than the other methods for removal of phenol.

The removal efficiency of this technique reached 97%. This corresponds to a residual phenol concentration of 3 $mg\,L^{-1}$. The equilibrium sorption was reached after 48 h. The least effective method for removal of phenol was the biodegradation using active bacteria alone [12].

12.3 SULFUR REMOVAL

The approaches for design, associated catalysis, and chemistry for deep desulfurization and deep dearomatization of hydrocarbon fuels, particularly diesel fuels, have been reviewed [13].

The challenge for the deep desulfurization of diesel fuels is the difficulty of removing the refractory sulfur compounds, particularly 4,6-dimethyldibenzothiophene, by a conventional hydrodesulfurization processes. The problem is exacerbated by the inhibiting effect of polyaromatics and nitrogen compounds, which exist in some diesel blend stocks on deep hydrodesulfurization.

A new design concept for sulfur-tolerant noble metal catalysts for low-temperature hydrogenation and a new desulfurization process of sulfur adsorption and capture under hydrogen have been described. Also biodesulfurization is a variant for desulfurization [13].

12.3.1 Refractory Sulfur Compounds

An adsorbent composition for removal of refractory sulfur compounds from refinery streams has been described [14]. The base component of the composition is spinel oxide; there is also a reactive metal oxide

Table 12.4 Composition for removing sulfur

Compound	**Amount (%)**	
Zinc nickel aluminate	50	45
Zinc nickel oxide	40	25
Rajasthan clay	10	15
Alumina	–	15.2
Properties		
BET surface area ($m^2\ g^{-1}$)	57	106
Pore volume ($cm^3\ g^{-1}$)	0.14	0.22
Average pore size (Å)	100	83

BET, Brunauer-Emmett-Teller.
Source: Kumar S, Pulikottil AC, Sharma A, Manna U, Santra M, Verma RP. Adsorbent composition for removal of refractory sulphur compounds from refinery streams and process thereof. US Patent 8 222 180, assigned to Indian Oil Corporation Limited (Faridabad, IN); 2012. http://www.freepatentsonline.com/8222180.html.

component and a bimetallic alloy as an adsorption enhancer component. The adsorption enhancer acts in synergy with the base component.

A process for the preparation of the composition has been described in detail [14]. An exemplary composition is shown in Table 12.4.

12.3.2 Refinery Off-Gas

The petroleum refining industry generates large quantities of low-value process gases, which typically contain high concentrations of sulfur compounds [15]. These refinery off-gas (ROG) streams are generated from various secondary processing technologies used in oil refining, such as catalytic cracking, hydrotreating, and delayed coking processes. The largest quantity of ROG streams is derived from petroleum cracking units.

ROG streams are composed of a wide variety of gases, including hydrogen, carbon monoxide, carbon dioxide, and saturated hydrocarbons and olefinic hydrocarbons.

The content of ethane and ethylene can be as high of 30% and the content of hydrogen is typically in the range of 15–50%. Sulfur compounds are mostly hydrogen sulfide, carbonyl sulfide, and organic sulfur compounds such as mercaptans, thiophenes, and sulfides. The concentration of H_2S can be greater than 1% by volume and the concentration of organic sulfur compounds can be several hundred parts per million.

Because of the lack of effective technologies for converting ROG streams into more valuable products or useful feed streams, many of these gas streams are used for their fuel value or, in many cases, simply flared. However, this can result in the emission of toxic or environmentally undesirable gases [15].

The conversion of high-sulfur ROG streams into more valuable low-sulfur hydrocarbon and hydrogen-containing streams can reduce energy losses, provide valuable feed streams for further processing, and eliminate many of the environmental concerns associated with the combustion of high-sulfur ROG streams.

Since many hydrocarbon conversion processes use expensive metal catalysts, the sulfur concentration must be lowered to avoid poisoning the metal catalysts.

Many refineries use an amine technology for sulfur removal. "Amine sulfur removal technology" refers to a group of processes that use aqueous solutions of various amine compounds to remove H_2S and CO_2 from sulfur-containing gases. The most commonly used amines are monoethanolamine, diethanolamine, and methyldiethanolamine.

These amine systems are very effective in removing H_2S, but are less effective in removing organic sulfur species such as mercaptans, thiophenes, sulfides, and other complex sulfur compounds.

For the removal of organic sulfur compounds, a caustic removal system is needed; however, caustic removal systems are expensive, moreover they use caustic reagents such as potassium hydroxide, which is considered toxic [15].

Another desulfurization method for fuel gas streams that contain organic sulfur compounds is a two-step process called hydrodesulfurization. This is the conversion of organic sulfur compounds into H_2S with the subsequent removal of the H_2S with an amine-based system or a solid sulfur adsorbent such as ZnO. This approach is typically used for the desulfurization of natural gas feedstocks and ROG streams having low sulfur levels (eg, 5–10 ppm) such as when natural gas is used as a feedstock in a steam methane reformer for the production of hydrogen.

Conventional hydrotreaters in steam methane reformer-based hydrogen plants operate at about 300–400°C and utilize waste heat from the steam methane reforming plant to preheat the feed to the hydrotreater. The catalyst used in conventional hydrotreaters is typically a CoMo or NiMo catalyst [15].

Also, the organic sulfur compounds in the ROG streams can be first hydrogenated in a hydrotreating process to form H_2S and then subsequently removed with conventional amine sulfur removal systems. However, for efficient hydrotreating of organic sulfur compounds, heat must be supplied and removed both economically and reliably for the system to convert organic sulfur into H_2S. Because the ROG streams are typically received at low pressures, such as 5–10 bar, the hydrotreater must be operated at elevated temperatures in the range of 290–370°C to ensure complete conversion of the organic sulfur species. Controlling the temperature within the hydrotreater becomes key for finding a cost-effective sulfur removal process, because waste heat is not always available [15].

Achieving the high temperature needed for hydrogenation of organic sulfur compounds without use of an external heat source can be a problem. Hydrogenation of gas streams containing olefins is an exothermic reaction, thereby providing heat to the reaction. If the ROG stream does not contain a sufficient concentration of olefins, the hydrogenation system will not be able to maintain the proper temperature for conversion of the organic sulfur compounds to H_2S and external heat must be provided to the reactor. If the ROG stream contains too high a concentration of olefins, the hydrotreating unit can overheat, causing the catalyst to be damaged or destroyed.

One solution to this problem is to dilute the high olefin concentration-containing ROG stream with a recycle stream from the hydrotreater product. This technique requires a recycle compressor, which complicates the system.

Another problem is that ROG streams usually have a significant variability in composition. This makes a hydrotreater with a recycle compressor-based system difficult to control [15].

A sulfur processing system that is flexible enough to process ROG streams having varying sulfur concentrations, varying organic sulfur compounds, and varying olefin content has been developed [15].

Organic sulfur compounds in ROG streams with either high or low concentrations of olefins can be converted into hydrogen sulfides. These sulfides can then be removed by conventional amine treating systems.

A process has been described that uses a catalytic reactor, with or without a hydrotreater, depending on the olefin concentration of the off-gas stream. The catalytic reactor may operate either in a hydrogenation mode or in an oxidation mode to convert most of the organic sulfur compounds into hydrogen sulfides. The design and construction of such a catalytic reactor has been described in detail [16]. In detail the process consists in [15]:

1. removing at least a portion of the H_2S from the feed stream by passing the feed stream through an amine absorber to produce an H_2S-depleted stream;

2. feeding a first portion of the H_2S-depleted stream into a catalytic reactor with the addition of oxygen to produce a hot effluent stream exiting the catalytic reactor at 340–450°C;
3. feeding a second portion of the H_2S-depleted stream into the hot effluent stream exiting the catalytic reactor to form a preheated combined stream, wherein the first portion and second portion are mixed in quantities such that the combined stream is fed into a hydrotreater to maintain the temperature of the hydrotreater at 340–450°C at the pressure used;
4. converting most of the organic sulfur compounds to hydrogen sulfide in the hydrotreater;
5. cooling the product gas stream exiting the hydrotreater;
6. feeding the cooled product gas stream to an amine sulfur-removal system to remove the H_2S and produce a hydrocarbon product stream.

A schematic view of the catalytic reactor is shown in Fig. 12.2.

When organic sulfur compounds such as mercaptans, thiophenes, and disulfides are present in combination with H_2S, the feed stream (3) is fed to the absorber (12) of the amine sulfur-removal system (ie, the amine treater), which reduces the H_2S concentration to less than about 30 ppm in the exiting H_2S-depleted stream (5). The absorber (12) may already be associated with or integrated into the amine treater, but will typically be a separate unit added for the propose of conducting the inventive process. Since most organic sulfur compounds cannot be substantially removed in the amine treater and may remain in amounts of greater than 50 ppm, additional processing is required [15].

The H_2S-depleted stream (5) leaving the absorber (12) and continuing to have high concentrations of organic sulfur compounds is first preheated in a recuperator (11) or other suitable heat exchanger, and then split into two streams (15 and 6). The first split stream (15) is sent to the catalytic reactor (2) and the second split stream (6) is sent to the hydrotreater (4). A sufficient amount of oxygen is introduced into the first split stream (15) through a line (16) before it passes into the catalytic reactor (2) to operate the reactor in the oxidation mode and to provide the heat needed for the conversion reaction. Optionally, the first split stream (15) may be preheated in a conventional start-up heater (14) to be heated during start-up. The flow of oxygen through the line (16) and the first split stream (15) are adjusted to provide sufficient reaction with the hydrogen present in the ROG feed stream (3) to produce water and heat to raise the temperature within the catalytic reactor (2) and the effluent stream (7) leaving the catalytic reactor

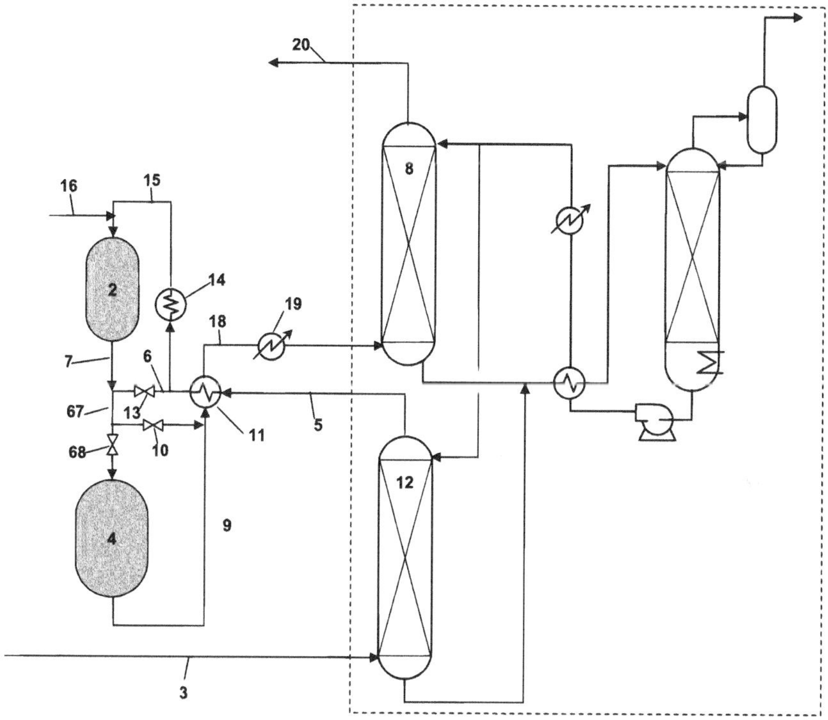

Figure 12.2 Catalytic reactor. *(Source: Shah MM, Drnevich RF, Papavassiliou V. Process for sulfur removal from refinery off gas. US Patent 8 808 654, assigned to Praxair Technology, Inc. (Danbury, CT); 2014. http://www.freepatentsonline.com/8808654.html).*

(2) to between 340 and 450°C. Because the reaction temperature of the effluent stream (7) is maintained above about 340°C, most of the organic sulfur compounds, typically more than 60%, and preferably greater than 70%, are converted to H_2S.

The effluent stream (7) exiting the catalytic reactor (2) is then mixed with the second split stream (6) and is fed through the line (67) to the hydrotreater (4). The mixing of the hot effluent stream (7) exiting the catalytic reactor (2) and the cooler second split stream (6) results in a preheated combined stream (67) with sufficiently high temperature so that the hydrotreater (4) can convert the organic sulfur compounds to H_2S, even with the low concentration of olefin present. The temperature of the combined stream (67) is controlled to be between 225 and 275°C by the volume of gas from the second split stream (6) added to the effluent stream (7). Depending on the olefin concentration of the feed stream (3), the volume of gas from the first split stream (15), the volume of gas from the second split stream (6), and the volume of oxygen added to the first split stream (15) can be adjusted to maintain the desired temperature range at the hydrotreater (4) entrance. This can be determined by one skilled in the art by monitoring the temperatures of the various steams or can be automated by use of processors and value actuators. The hydrotreated effluent stream (9) exits the hydrotreater (4) at about 340–400°C, is cooled by heat exchange with the H_2S-depleted stream (5) through the recuperator (11) and is sent through the line (18) to the cooler (19) and to the absorber (8) of the amine treater to remove H_2S. The hydrocarbon product stream (20) will have a low level of sulfur remaining, preferably below 20 ppm sulfur compounds. The absorber (8) is typically already present in the existing amine treater. Further processing can be done to this stream if desired to remove sulfur with a solid sulfur adsorbent such as zinc oxide, iron oxide, or activated carbon or by caustic treatment or any other polishing sulfur removal technique to further reduce sulfur levels by means of conventional systems. The amine treating system is shown in Fig. 12.2 within the doted box [15].

12.4 SELENIUM REMOVAL

Selenium compounds have been reported to constitute 0.9 ppm of Earth's crust [17]. Selenium is important as a trace mineral used to make the enzyme glutathione peroxidase, which is involved in fat metabolism and therefore is found in many living organisms. In addition, it is commonly found in various amounts in crude oil, coal, and other fossil fuels originating from decomposed organic matter or leached out of nearby minerals. Selenium compounds are also found naturally in groundwaters and in agricultural runoffs from the use of selenium-containing insecticides and herbicides.

Unfortunately, selenium is highly toxic and it can cause harm even in small quantities. Harmful effects include dermatitis, central nervous system disturbance, nephrosis, hemorrhagic necrosis of the pancreas and adrenal cortex, and when in large enough doses, death [17].

Therefore, regulations have been formulated that have limited the permissible amount of selenium in domestic supplies of water to 10 ppb. As a result, wastewater produced from activity involving selenium-containing materials is difficult to dispose of. In addition, because of the toxicity of selenium, even these strict standards may be uniformly further limited in the future.

Unfortunately, the chemical properties of selenium make its removal from solutions difficult [17]. Although insoluble in its elemental state, selenium has four oxidation states (−2, +2, +4, and +6), which allows it to readily form a number of salts that are highly soluble and therefore very hard to remove from solution [18].

A method for selenium removal involves the use of a ferric salt and eventual pH adjustment [19]; however, the effectiveness is less than 70%. Another method uses poly(dithiocarbamate) derivatives [20].

More recently, a method for removing selenium from a liquid has been described. The method consists in [17]:

1. adding an oxidant to the liquid,
2. adjusting the pH of the liquid to below 7.5,
3. adding as much ferric salt such that less than a quarter of selenium in the liquid precipitates, and
4. adding dithiocarbamate or a poly(dithiocarbamate).

Typically refinery wastewater has a pH of more than 7 and thus its pH needs to be adjusted lower through the addition of acid. Experiments confirmed that the most effective application of this process occurs in a pH range of 6–7. Selenite easily forms complexes with the ferric ion and poly(dithiocarbamate). The aggregates form large flocculates, which can be separated by a solid-liquid separation method [17].

12.5 NICKEL REMOVAL

The aerial deposition of Ni from a refinery in Port Colborne, Ontario, Canada, has resulted in the enrichment of 29 km^2 of land with Ni concentrations exceeding the Canadian Ministry of the Environment's remedial action level of 200 $mg\,kg^{-1}$ [21].

Several studies on these soils have shown that making the soils calcareous was effective in reducing chemically extractable Ni, as well as alleviating Ni phytotoxicity symptoms in vegetable crops that had been grown near the refinery. Conversely, the addition of dolomitic limestone resulted in an increased uptake of Ni by the Ni hyperaccumulator *Alyssum murale* "Kotodesh"—a plant whose use was proposed as a remediation strategy for the area.

Multiple techniques have been used to assess directly the roles that the soil type and lime treatments play in altering the speciation of Ni in the

Welland loam and quarry muck soils around the refinery and relate these findings to the mobility and bioavailability of Ni [21].

Stirred-flow dissolution experiments using pH 4 HNO_3 showed that Ni release from the limed quarry muck and Welland loam soils was reduced by some 0.1% relative to the unlimed soils (approximately 2.0%). Electron microprobe analysis identified approximately spherical NiO and Ni particles, which are associated with no other metals, and ranged from 5 to 50 μm in diameter.

Synchrotron micro X-ray absorption fine structure spectroscopy and X-ray fluorescence spectroscopy showed that Ni and Al layered double hydroxide phases were present in both the limed and the unlimed mineral soils, with a tendency toward stabler Ni species in the limed soil. This is possibly aided by the solubilization of Si with increasing pH. In the muck soils, Ni organic complexes (ie, fulvic acid) dominate the speciation in both limed and unlimed soils.

It has been shown that both the soil type and the treatment method have a pronounced effect on the speciation of Ni in the soils surrounding the refinery.

There is evidence that layered double hydroxide Ni-Al phases can form in anthropogenically enriched mineral field soils at circumneutral pH, and can lead to a reduction in Ni mobility.

In organic soils, Ni is strongly complexed by the organic matter of soil, and this is enhanced with liming. An increased accumulation of Ni by *Alyssum murale* "Kotodesh" grown in the limed muck and loam soils indicates that the plant may be capable of removing Ni from those fractions that are typically considered unavailable to most plants [21].

12.5.1 Phosphorus Gelling Agents

Use of commonly used phosphoric acid esters may suffer from the problem that they decompose in refinery distillation towers to form volatile phosphorus, which condenses on the trays of the distillation towers and causes plugging. Also, the phosphoric acid ester may itself be volatile, dependent on its molecular weight. Improved methods and compositions for fracturing subterranean formations with gelled liquid hydrocarbon fracturing fluids have been described. The gelling agent is a ferric iron or aluminum polyvalent metal salt of a phosphoric acid ester with substituents that prevent the volatilization and degradation of the material [22].

Liquid hydrocarbons can be gelled by the introduction of a phosphate ester and a cross-linking agent for use in oil recovery. Generally a

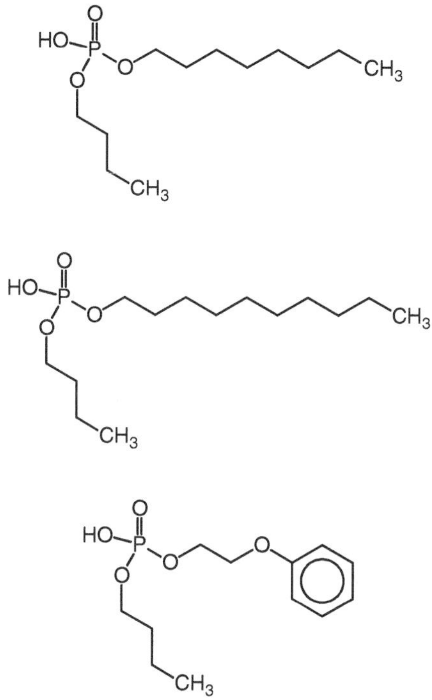

Figure 12.3 Phosphate esters. *(Source: Delgado E, Keown B. Low volatile phosphorous gelling agent. US Patent 7 622 054, assigned to Ethox Chemicals LLC (Greenville, SC); 2009. http://www.freepatentsonline.com/7622054.html).*

proppant, delayed gel breaker, and other modifiers are added to the gelled hydrocarbon to improve the oil recovery process. The phosphate ester used in this method and composition for oil recovery shows low volatility and good stability during hydrocarbon distillation, resulting in improved distillation efficiency in a refinery [22]. Phosphate esters are shown in Fig. 12.3.

12.6 SPENT REFINERY CATALYSTS

The effect of sequential leaching of metals in spent catalysts has been tested. As sequential leaching techniques, the following were used [23]:

1. bioleaching followed by bioleaching and
2. bioleaching followed by chemical leaching

The dissolution of Mo, Ni, V, and Al from a differently pretreated (acetone washed and decoked) spent catalyst was measured. X-ray photoelectron spectroscopy characterization of the spent catalyst samples suggested the presence of the metals in their oxide and sulfide forms.

Bioleaching followed by bioleaching with either *Acidithiobacillus thiooxidans* or *Acidithiobacillus ferrooxidans* significantly enhanced the removal of Al, Ni, and V from an acetone-washed spent catalyst in comparison with a decoked spent catalyst.

In contrast, bioleaching using *Acidithiobacillus thiooxidans* or *Acidithiobacillus ferrooxidans* followed by alkali leaching remarkably enhanced the removal of Mo. Bioleaching using *Acidithiobacillus thiooxidans* followed by alkali leaching is an optimum strategy yielding a maximum of 96% Mo in 125 h [23].

Bioleaching also affects the fractionation of the metals remaining in the treated spent catalyst. At the end of the process, most of the metals remaining in the spent catalyst were found in the stabler fractions, ensuring the safe disposal of spent catalyst [24].

12.7 WATER NETWORKS

The procedures for the design and retrofitting of water networks for refineries have been reviewed [25]. Emphasis has been given to showing the results for refineries; however, the methods are valid for any process plants.

It has been shown that the problem can be decomposed into the design of two interacting subsystems. One problem is the freshwater and wastewater reuse allocation and the other is the wastewater treatment problem.

It is also outlined how the wastewater treatment problem can be modeled as a distributed and decentralized treatment [25]. The road map toward zero liquid discharge and energy-integrated solutions has been discussed.

Several approaches have been discussed and the methods of programming have been shown. Mathematical programming can produce globally optimal solutions and practically important suboptimal solutions when conceptual insights are used to build the models [25].

12.8 SPECIAL ASPECTS

12.8.1 Risks for Workers

A dynamic retrospective study was performed to examine the mortality experience of workers in a refinery [26]. For this, 8666 regular employees who worked for at least 1 month during the period from January 1, 1970 through December 31, 1977 were checked as were retirees.

The mortality from all causes of death was lower than expected when compared with that of the US population of similar age, sex, and race. The analysis of mortality by specific site of cancer revealed elevated standardized mortality ratios for cancer of the kidney, testis, brain, central nervous system, pancreas, and lymphopoietic sites; however, none of these elevations were statistically significant [26]. An updated study, some 10 years after the first study, showed virtually similar results [27].

Another study revealed that the overall mortality and cancer mortality for both refinery and chemical employees were quite favorable compared with those for the residents in the local population. For both refinery and chemical plant employees, the mortality rate due to cancers of all lymphopoietic tissue increased with increasing duration of employment [28]. Another study came to similar conclusions [29].

12.8.2 Toxicities

Studies have been performed to determine whether the systemic and developmental toxicity can be related to the presence and the levels of certain classes of refinery stream components [30]. Systemic toxicology studies were performed in the rat on 13 refinery streams, listed in Table 12.5.

Rats were exposed daily via repeated dermal administration at several doses [30]. Developmental toxicology studies were performed on these same streams with the following exceptions: only two coker heavy gas oils were tested and visbreaker gas oil was not tested. At the end of the 13-week

Table 12.5 Refinery streams for toxicology studies

Stream	
Clarified slurry oil	Coker light gas oil
Distillate aromatic extract	Heavy atmospheric gas oil
Coker heavy gas oil	Heavy vacuum gas oil
Light catalytically cracked naphtha	Light cycle oil
Syn-tower bottoms	Vacuum tower overhead
Visbreaker gas oil	

Source: Feuston MH, Low LK, Hamilton CAE, Mackerer CR. Correlation of systemic and developmental toxicities with chemical component classes of refinery streams. Toxicol Sci 1994;22(4):622–30. http://dx.doi.org/10.1093/toxsci/22.4.622.

studies skin irritation, body and organ weights, aberrant hematological findings, and serum chemistry were noted.

For developmental toxicity studies some of these same end properties were considered, but also resorption and change in feal body weight were included [30]. The toxicity was correlated with concentrations of polycyclic aromatic compounds composed of three, four, five, six, and seven rings. These decreased thymus weight, increased liver weight, and resulted in aberrant hematological and serum chemistry findings.

Polycyclic aromatic compounds with nonbasic nitrogen heteroatoms increased mortality, decreased body weight, decreased thymus weight, increased liver weight, and decreased the hemoglobin content and hematocrit; further, they decreased fetal body weight. Polycyclic aromatic compounds containing sulfur heteroatoms decreased red blood cell and platelet counts and increased sorbitol dehydrogenase levels.

A relationship between two-ring polycyclic aromatic compounds and skin irritation was demonstrated. The severity of this effect was ranked against the concentration of the component class and the statistical significance determined by Spearman rank-order correlation.

In summary, the presence and the severity of systemic and developmental toxicity were dependent on the levels of the polycyclic aromatic compounds [30]. These classes of compounds are believed to be the major contributors to the dermal carcinogenic activity of crude oils, certain refinery streams, and poorly refined lube oils.

REFERENCES

[1] Hoyt D, Raun LH, Measured and estimated benzene and volatile organic carbon (VOC) emissions at a major U.S. refinery/chemical plant: comparison and prioritization. J Air Waste Manage Assoc 2015;65(8):1020–31. http://dx.doi.org/10.1080/10962247.2015.1058304.

[2] Maher PJ. Method for treating refinery waste streams in a fluid catalytic cracking unit and an improved catalytic cracking unit for processing refinery waste streams. US Patent 8 114 272, assigned to ExxonMobil Research & Engineering Company (Annandale, NJ); 2012. http://www.freepatentsonline.com/8114272.html.

[3] Zhong J, Sun X, Wang C, Treatment of oily wastewater produced from refinery processes using flocculation and ceramic membrane filtration. Sep Purif Technol 2003;32(1–3):93–8. Seventh International Conference on Inorganic Membranes, http://www.sciencedirect.com/science/article/pii/S1383586603000674, http://dx.doi.org/10.1016/S1383-5866(03)00067-4.

[4] Hami ML, Al-Hashimi MA, Al-Doori MM, Effect of activated carbon on {BOD} and {COD} removal in a dissolved air flotation unit treating refinery wastewater. Desalination 2007;216(1–3):116–22. http://www.sciencedirect.com/science/article/pii/S0011916407004298, http://dx.doi.org/10.1016/j.desal.2007.01.003.

[5] Sommese AG, Sivakumar A. Polymers containing vinylamine/vinylformamide as demulsifiers in oily wastewaters. US Patent 5 702 613, assigned to Nalco Chemical Company (Naperville, IL); 1997. http://www.freepatentsonline.com/5702613.html.
[6] Kong J, Li K, Oil removal from oil-in-water emulsions using {PVDF} membranes. Sep Purif Technol 1999;16(1):83–93. http://www.sciencedirect.com/science/article/pii/S1383586698001142, http://dx.doi.org/10.1016/S1383-5866(98)00114-2.
[7] Tomaszewska M, Preparation and properties of flat-sheet membranes from poly(vinylidene fluoride) for membrane distillation. Desalination 1996;104(1–2):1–11. http://dx.doi.org/10.1016/0011-9164(96)00020-3.
[8] Deshmukh SP, Li K, Effect of ethanol composition in water coagulation bath on morphology of {PVDF} hollow fibre membranes. J Membr Sci 1998;150(1):75–85. http://www.sciencedirect.com/science/article/pii/S0376738898001963, http://dx.doi.org/10.1016/S0376-7388(98)00196-3.
[9] Jian K, Pintauro PN, Asymmetric {PVDF} hollow-fiber membranes for organic/water pervaporation separations. J Membr Sci 1997;135(1):41–53. http://www.sciencedirect.com/science/article/pii/S0376738897001178, http://dx.doi.org/10.1016/S0376-7388(97)00117-8.
[10] Zularisam AW, Ismail AF, Salim R, Behaviours of natural organic matter in membrane filtration for surface water treatment—a review. Desalination 2006;194(1–3):211–31. http://dx.doi.org/10.1016/j.desal.2005.10.030.
[11] Yuliwati E, Ismail AF, Effect of additives concentration on the surface properties and performance of PVDF ultrafiltration membranes for refinery produced wastewater treatment. Desalination 2011;273(1):226–34. http://dx.doi.org/10.1016/j.desal.2010.11.023.
[12] Altaher H, Ashour I, Sawalha F, Maraqah M, A comparative study for the treatment of refinery synthetic wastewater containing phenol. Sci Technol Dev 2015;34(1):16–26. http://dx.doi.org/10.3923/std.2015.16.26.
[13] Song C, Ma X, New design approaches to ultra-clean diesel fuels by deep desulfurization and deep dearomatization. Appl Catal B Environ 2003;41(1–2):207–38. 3rd International Conference on Environmental Catalysis, http://www.sciencedirect.com/science/article/pii/S0926337302002126, http://dx.doi.org/10.1016/S0926-3373(02)00212-6.
[14] Kumar S, Pulikottil AC, Sharma A, Manna U, Santra M, Verma RP. Adsorbent composition for removal of refractory sulphur compounds from refinery streams and process thereof. US Patent 8 222 180, assigned to Indian Oil Corporation Limited (Faridabad, IN); 2012. http://www.freepatentsonline.com/8222180.html.
[15] Shah MM, Drnevich RF, Papavassiliou V. Process for sulfur removal from refinery off gas. US Patent 8 808 654, assigned to Praxair Technology, Inc. (Danbury, CT); 2014. http://www.freepatentsonline.com/8808654.html.
[16] Drnevich RF, Papavassiliou V. Steam methane reforming method. US Patent 7 037 485, assigned to Praxair Technology, Inc. (Danbury, CT); 2006. http://www.freepatentsonline.com/7037485.html.
[17] Schwarz DE, Shah JT. Removal of selenium from refinery wastewater. US Patent 8 282 835, assigned to Nalco Company (Naperville, IL); 2012. http://www.freepatentsonline.com/8282835.html.
[18] Kapoor A, Tanjore S, Viraraghavan T, Removal of selenium from water and wastewater. Int J Environ Stud 1995;49(2):137–47. http://dx.doi.org/10.1080/00207239508711016.
[19] Alexander R, Lord CJ, Mitchell SC. Selenium removal from water. US Patent 7 419 602, assigned to ConocoPhillips Company (Houston, TX); 2008. http://www.freepatentsonline.com/7419602.html.

[20] Miller TM, Goodman WH. Removal of selenium from water by complexation with polymeric dithiocarbamates. US Patent 5 510 040, assigned to Nalco Chemical Company (Naperville, IL); 1996. http://www.freepatentsonline.com/5510040.html.
[21] McNear Jr DH, Chaney RL, Sparks DL, The effects of soil type and chemical treatment on nickel speciation in refinery enriched soils: a multi-technique investigation. Geochim Cosmochim Acta 2007;71(9):2190–208. http://www.sciencedirect.com/science/article/pii/S0016703707000786, http://dx.doi.org/10.1016/j.gca.2007.02.006.
[22] Delgado E, Keown B. Low volatile phosphorous gelling agent. US Patent 7 622 054, assigned to Ethox Chemicals LLC (Greenville, SC); 2009. http://www.freepatentsonline.com/7622054.html.
[23] Srichandan H, Pathak A, Singh S, Blight K, Kim DJ, Lee SW, Sequential leaching of metals from spent refinery catalyst in bioleaching/bioleaching and bioleaching/chemical leaching reactor: comparative study. Hydrometallurgy 2014;150:130–43. http://www.sciencedirect.com/science/article/pii/S0304386X14002163, http://dx.doi.org/10.1016/j.hydromet.2014.09.019.
[24] Pathak A, Srichandan H, Kim DJ, Fractionation behavior of metals (0200003*D*013*C*013*EAl*, 0200003*D*013*C*013*ENi*, 0200003*D*013*C*013*EV*, and 0200003*D*013*C*013*EMo*) during bioleaching and chemical leaching of spent petroleum refinery catalyst. Water Air Soil Pollut 2014;225(3):1893. http://dx.doi.org/10.1007/s11270-014-1893-1.
[25] Bagajewicz M, A review of recent design procedures for water networks in refineries and process plants. Comput Chem Eng 2000;24(9–10):2093–113. http://www.sciencedirect.com/science/article/pii/S0098135400005792, http://dx.doi.org/10.1016/S0098-1354(00)00579-2.
[26] Hanis NM, Holmes TM, Shallenberger LG, Jones KE, Epidemiologic study of refinery and chemical plant workers. J Occup Environ Med 1982;24(3). http://journals.lww.com/joem/abstract/1982/03000/epidemiologic_study_of_refinery_and_chemical_plant.12.aspx.
[27] Shallenberger LG, Acquavella JF, Donaleski D, An updated mortality study of workers in three major united states refineries and chemical plants. Occup Environ Med 1992;49(5):345–54. http://dx.doi.org/10.1136/oem.49.5.345.
[28] Tsai SP, Gilstrap EL, Cowles SR, Snyder PJ, Ross CE, Long-term follow-up mortality study of petroleum refinery and chemical plant employees. Am J Ind Med 1996;29(1):75–87. http://dx.doi.org/10.1002/(SICI)1097-0274(199601)29:1<75::AID-AJIM10>3.0.CO;2-X.
[29] Marsh GM, Enterline PE, McCraw D, Mortality patterns among petroleum refinery and chemical plant workers. Am J Ind Med 1991;19(1):29–42. http://dx.doi.org/10.1002/ajim.4700190105.
[30] Feuston MH, Low LK, Hamilton CAE, Mackerer CR, Correlation of systemic and developmental toxicities with chemical component classes of refinery streams. Toxicol Sci 1994;22(4):622–30. http://dx.doi.org/10.1093/toxsci/22.4.622.

INDEX

Note: Page numbers followed by *f* indicate figures and *t* indicate tables.

E

F

G

H

R

FSC
www.fsc.org
MIX
Papier | Fördert
gute Waldnutzung
FSC® C083411

Kolibri 360 GmbH
Ferdinand-Jühlke-Straße 7
99095 Erfurt, Deutschland
produktsicherheit@kolibri360.de